Refrigeration Technician's Pocket Book

Refrigeration Technician's Pocket Book

F. H. MEREDITH,
MIHVE, FInstR

Senior Lecturer, South East London College
Training Consultant, World Health Organization

Butterworths
London Boston Sydney Wellington Durban Toronto

First published 1981
Reprinted 1981, 1985

British Library Cataloguing in Publication Data

Meredith, F H
Refrigeration technician's pocketbook.
1. Refrigeration and refrigerating machinery – Maintenance and repair
I. Title
621.5′7 TP 492 80-41068

ISBN 0-408-00545-9

Typeset by Reproduction Drawings Ltd, Sutton, Surrey
Printed in England by The Thetford Press Limited, Thetford, Norfolk

PREFACE

This Pocket Book has been written to assist all those who are involved in the installation or servicing of refrigeration plant.

Although construction and operation of refrigeration plant are adequately covered by a number of well-known works, these are of limited help when trying to find and repair a fault, or install and commission a plant. This book covers these practical problems.

It describes in simple, practical terms the faults which commonly occur and how to recognise and rectify them. Similarly it explains, in an easy-to-follow way, the major problems which arise when installing a refrigeration plant.

As a Pocket Book, it is intended to be used 'on the job' and has been written with this in mind. It adopts a step-by-step approach and gives simple clear instructions of what to do at each stage. The written text is supported by simplified drawings and diagrams which the user will find easy to understand when working in the field.

The experienced service engineer will find this book useful as an 'aide-memoire', particularly when faced with an unusual problem. The student and trainee engineer will find it invaluable as a bridge between classroom theory and the understanding that is necessary to carry out repair and installation work unaided, with both confidence and success.

The author gratefully acknowledges the help of the following: *Feedback Instruments Ltd*. The use of their training facilities, particularly the RAC 192 trainer, was of great help in evaluating the fault finding sections of the book. *Mr P. S. Balm, Chief Engineer, Electrorentals Ltd*. He not only read, and commented in detail on the text but also ensured that the subject matter truly reflected the needs of the Industry.

FRANK H. MEREDITH

CONTENTS

SECTION A

VAPOUR COMPRESSION TYPE REFRIGERATION SYSTEMS

1 OPERATING PRINCIPLES

The vapour compression type plant operates on the principle of variation of boiling point with change in pressure.

If the pressure of any liquid is increased, its boiling point will be raised. Conversely, if the pressure of the liquid is reduced, its boiling point will be lowered. For example, water at normal atmospheric pressure (14.7 lb/in^2) boils at 212 °F (100 °C). If its pressure is increased to 20 lb/in^2, the boiling point is raised to approximately 228 °F(109 °C). If its pressure is reduced to 10 lb/in^2, the boiling point is lowered to approximately 193 °F (89 °C).

All liquids behave in this way and the 'working liquid' of a refrigeration plant is continually having its pressure changed, in order to vary its boiling point.

When the working liquid (referred to as a REFRIGERANT) has its pressure reduced, by the suction effect of a compressor, the boiling point is lowered. The low temperature at which the liquid evaporates (boils) enables cooling to be carried out.

When the refrigerant, which has evaporated to form a vapour, is compressed, the boiling point is raised. If it is now cooled BELOW the boiling point, the refrigerant vapour will condense to re-form as liquid, ready for further evaporation and cooling.

The complete cycle is shown in Fig. 1. The sequence of operations is as follows:

(a) Liquid is admitted to a coil of pipe (or similar), called the 'evaporator', by a control valve.

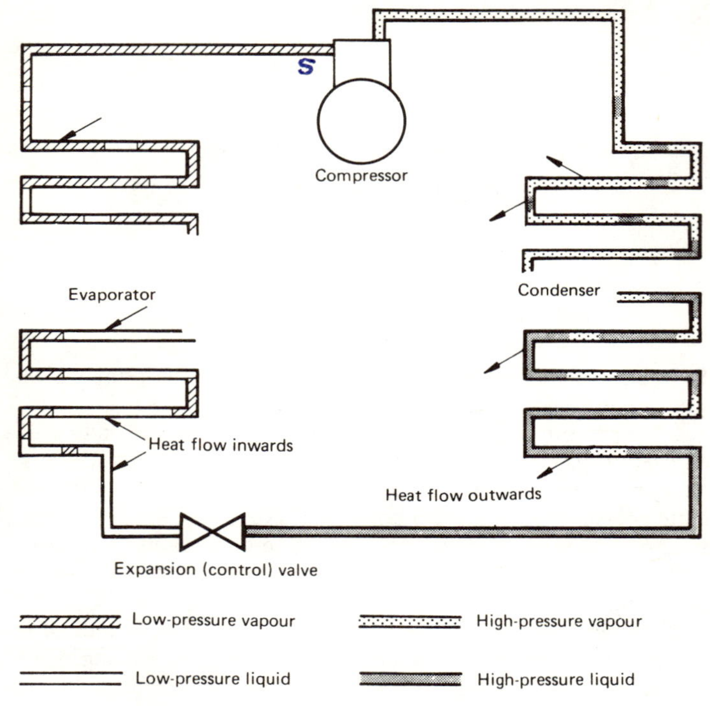

Figure 1. Vapour compression refrigeration cycle

(b) The coil of pipe is at low pressure so that the liquid evaporates (boils) at low temperature and produces a cooling effect.
(c) The outlet of the pipe coil is connected to the suction side of a compressor which reduces the pressure in the coil and draws off the vapour formed by the liquid evaporating.
(d) The compressor delivers the refrigerant at high pressure, and therefore with a high boiling point, into another pipe coil, the condenser.
(e) The outside of this second coil of pipe is cooled, with air or water, so that the hot, high pressure vapour condenses to form liquid and complete the cycle.

The operating sequence, (a) to (e) above, will now be repeated. This time the action of each component will be fully described and its correct technical term used.

(a) The control of liquid flow is carried out by the EXPANSION VALVE. It is supplying high pressure liquid to a low pressure coil, so that an expansion process occurs. The expansion valve has two purposes:
(i) To maintain, in conjunction with the compressor, the essential low pressure in the evaporating, and high pressure in the condensing, sides of the system.
(ii) To regulate the flow of liquid refrigerant so that maximum cooling can be obtained, without allowing unevaporated liquid to be drawn into the compressor suction.

(b) The pipe coil into which the liquid is fed by the expansion valve is called the EVAPORATOR. It is here that the liquid evaporates and, in changing from liquid to vapour, takes its latent heat from the space to be cooled. The evaporator will always be at low pressure and temperature with the flow of heat inwards. (Heat is added to the refrigerant).

(c) The COMPRESSOR, like the expansion valve, has two purposes. It must first reduce the pressure in the evaporator until the liquid refrigerant evaporates at the correct, low, temperature (and maintain the reduction by drawing off the vapour formed by the evaporation of the liquid, allowing further evaporation to occur).

(d) The second action of the COMPRESSOR is to compress the vapour, in order to raise its boiling point. To supply the expansion valve with liquid the vapour delivered by the compressor MUST be condensed. This will only occur if the boiling point is ABOVE that of the cooling air or water.

(e) The CONDENSER receives the hot, high pressure vapour from the compressor and cools it. The flow of heat is now outwards (the heat added to the refrigerant in the evaporator is rejected in the condenser)

Condensation occurs as the refrigerant passes through the condenser and high pressure liquid is formed to feed the expansion valve and complete the cycle.

From the explanations given it will be understood that this system of refrigeration depends upon

maintaining a low pressure in the evaporator,
maintaining a high pressure in the condenser.

The essential pressure difference is maintained by the action of the compressor and it is for this reason that it is called the VAPOUR COMPRESSION CYCLE.

The evaporating pressure of an operating plant will be determined by the evaporating temperature required (for low temperature storage of food a low evaporating temperature and pressure is necessary).

The condensing pressure of a plant will be determined by the temperature of the cooling air or water (in order that heat can flow from the condensing refrigerant the air or water MUST be at a lower temperature than the refrigerant). If the cooling air temperature rises (on a hot day), the condensing temperature must rise to give the same temperature difference between the condensing refrigerant and the air.

The operating pressures of any plant are therefore NOT fixed but vary depending upon the required evaporating temperature and the temperature of the air or water used to cool the condenser.

To obtain the maximum cooling effect, efficient operation and a high degree of reliability (reduction of breakdown) refrigeration plants operating on the vapour compression principle should

(a) Evaporate at as high a pressure as possible (provided that the required cooling effect is being produced)
(b) Condense at as low a pressure as possible (provided that condensation to liquid is being obtained).

Normal operating conditions and the faults which cause abnormal pressures are described in SECTION K MECHANICAL FAULT FINDING.

2 COMMERCIAL REFRIGERATION SYSTEMS

Commercial systems are those used for applications such as food display units, large 'walk in' refrigerated cabinets, 'free standing' console type air conditioning units and pre-fabricated cold rooms.

They normally use refrigerants 12, 22 or 502 with semi-hermetic compressors and thermostatic type expansion valves. A typical layout of a commercial system is shown in Fig. 2.

The purpose of the components fitted to a plant of this type are as described below. It must be emphasied that not ALL commercial systems incorporate the range of components shown. This comprehensive list is intended to show the components that may be fitted and the reason for using them.

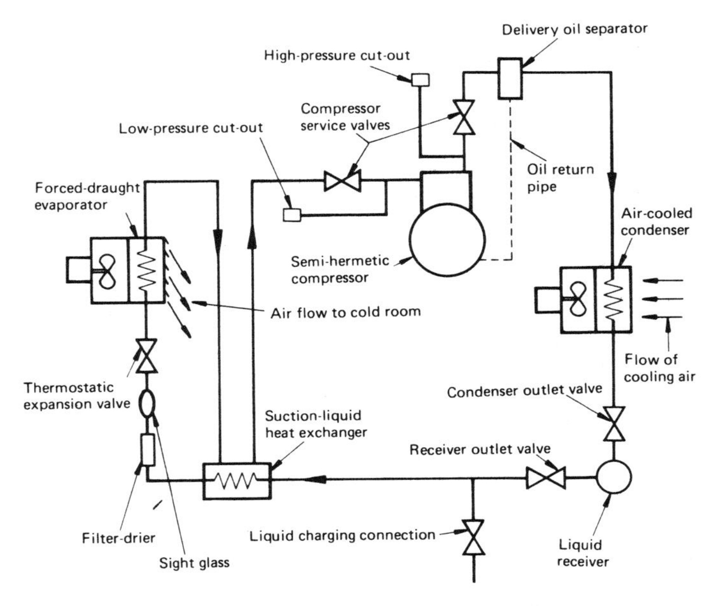

Figure 2. Layout of typical commercial refrigeration system

Filter Drier

Commercial type systems are easily contaminated with moisture (from the air) and small particles of dirt. These contaminants enter the system during servicing work and cause restriction of liquid flow (particularly at the expansion valve where the opening is small, moisture freezes and a complete blockage occurs).

The component fitted consists of a very fine filter, which will trap small particles of dirt, followed by a drying agent. Silica gel, activated alumina or molecular sieves are commonly used as drying agents. They all have a high capacity for absorbing moisture.

The filter-drier should be fitted immediately before the expansion valve, except where a sight glass is used. The order should then be filter-drier, sight glass, expansion valve.

Sight Glass

This component MUST be fitted immediately before the expansion valve. It has a glass top so that the flow of refrigerant liquid can be seen. The presence of vapour bubbles appearing in the sight glass indicates a fault in the system. Either

(a) There is a restriction between the condenser outlet and the sight glass. Common causes of restriction are blocked filter-driers, partially closed valves and damaged pipe.
(b) The system is undercharged. There is insufficient liquid to fill the pipe connecting to the condenser and vapour is passing along it. This is usually an indication that the system has developed a leak.

Compressor Service Valves

These are fitted to both the suction and delivery connections to the compressor.

They can be fully closed, by turning in a clockwise direction,

to isolate the compressor from the remainder of the system. This enables repair and maintenance work to be carried out without losing the charge of refrigerant from the system.
Note. The delivery service valve must NOT be closed while the compressor is still running. Unless there is an outlet for the compressed vapour damage may occur.

The service valves can also be fully opened, by turning in an anti-clockwise direction. This closes off the 'back seat' connections of the valves to which are fitted gauges (to determine the operating pressures), charging connections (for adding refrigerant vapour to the suction side of the compressor) and a purge connection (on the delivery valve, to remove excess refrigerant and any air or other non condensable gas).

Low Pressure Cut-out

This is a pressure-operated switch, fully described under SECTION F CONTROLS. Although temperature control of commercial plant is normally by thermostat, a low pressure cut out is usually fitted to avoid damage (to both the product and the plant) by low evaporating temperature if the thermostat should fail.

High Pressure Cut-out

This is a safety device intended to protect the plant and personnel from excessive condensing pressures. It is also fully described under SECTION F CONTROLS.
Note. Pressure cut-outs MUST be connected to the system at a point where they cannot be isolated by closing a valve or similar.

Delivery Oil Separator

With the type of refrigerants used for commercial applications oil is discharged from the compressor with the vapour and circulates around the system.

Although small quantities of oil are beneficial—they lubricate the expansion and other valves—large amounts are undesirable:

(a) The oil will coat the inside of the evaporator and restrict the heat flow rate.
(b) The suction pipe to the compressor must be arranged to 'carry back' the circulating oil and this is more difficult when large quantities are present.

A separating device is therefore fitted to the compressor delivery which removes the oil droplets from the refrigerant vapour (by centrifugal action or impingement on a series of baffles). The oil removed by the separator is returned, normally by a float valve, to the compressor crank-case.

Liquid Receiver and Associated Valves

A storage receiver is fitted to all systems in the commercial range. It is a high pressure cylindrical steel shell, usually mounted horizontally, and fitted with a fusible plug which will melt in the event of a fire occurring and release the refrigerant from the system before a dangerously high pressure is reached.

The receiver is intended to hold a reserve of refrigerant so that liquid is always available to supply the expansion valve when it opens (an increase of cooling load).

It is also used for storing the charge of refrigerant in the system while repairs are being carried out. The valves shown are provided for this purpose. The procedure for transferring refrigerant into the receiver is as follows:

(a) Close the receiver outlet valve.
(b) Run the compressor until the suction pressure is just ABOVE atmospheric.
(c) Switch off the compressor.
(d) Close the condenser outlet valve.

Liquid Charging Connection

Vapour charging, at the compressor suction, is slow and therefore large plants (which require greater quantities of refrigerant to be added) are fitted with the liquid charging connection shown.

Liquid occupies a smaller volume than vapour and can therefore be charged into the system more quickly. The pressure in the system is, however, greater than that in the charging cylinder. To add refrigerant through the liquid charging connection, it is therefore necessary to reduce the pressure in the system at this point by partially closing the receiver outlet valve.

Suction-liquid Heat Exchanger

This component uses the cold vapour leaving the evaporator to cool the warm liquid leaving the condenser, before it reaches the expansion valve. Heat exchangers of this type are used for one or more of the following reasons:

(a) To increase the cooling capacity of the plant. Warm liquid passing through the expansion valve has to be cooled which reduces the liquid's capacity to cool at the evaporator.
(b) To prevent vapour forming in liquid pipes, due to pressure drop caused by friction loss, or 'static lift' to a high level expansion valve. Vapour thus formed restricts the capacity of the expansion valve to supply liquid and also causes wear of the valve seat.
(c) To avoid insulating the compressor suction pipe. If this pipe carries cold vapour and is not insulated condensation will occur on the outside. This is troublesome since it drips onto the floor, particularly when the compressor shuts down. The warm vapour leaving a heat exchanger will not cause this problem.
(d) Small quantities of liquid are sometimes carried over with the suction vapour and damage the compressor. This problem is eliminated by the use of the suction-liquid heat exchanger.

Any liquid present in the suction vapour will immediately evaporate on coming into thermal contact with the warm liquid from the condenser.

Forced Draught Evaporators

Although some applications in the commercial range use tube evaporators, for cold room work, a positive circulation of air is essential. The type of evaporator normally used incorporates a fan mounted at the rear of a casing which holds the evaporator coils. The coils are finned externally, to give a larger effective cooling surface, and the cold room air is circulated across them.

The evaporator fan normally runs continuously, so that defrosting of the coils is achieved during the time that the compressor is shut down, provided that the cold room is at a temperature above freezing.

Air Cooled Condensers

These are used for most commercial applications to overcome the problems of providing, circulating and cooling condenser water.

Forced draught condensers are used, similar in construction and action to forced draught evaporators. They are sometimes used on the same bed plate as the compressor when they are referred to as 'condensing units'.

A good circulation of clean cool air is essential for satisfactory operation and for this reason, particularly in the larger sizes, they are frequently mounted at high level, remote from the compressor.

3 INDUSTRIAL REFRIGERATION SYSTEMS

The term 'Industrial' is used for refrigeration plants having a high cooling capacity. Although frequently used for industrial work, such as food processing, they cover a wide range of other applications where a high rate of heat extraction is required. Typical examples of industrial refrigerating plant are large purpose-built cold stores and ice-making installations.

They frequently use AMMONIA as the refrigerant with open type compressors. The difference between commercial and industrial cannot be defined in terms of cooling capacity or equipment used. An industrial system is, however, generally accepted as one which requires an attendant on duty, whereas commercial plant is normally fully automatic.

A typical industrial system is shown in Fig. 3, where ammonia

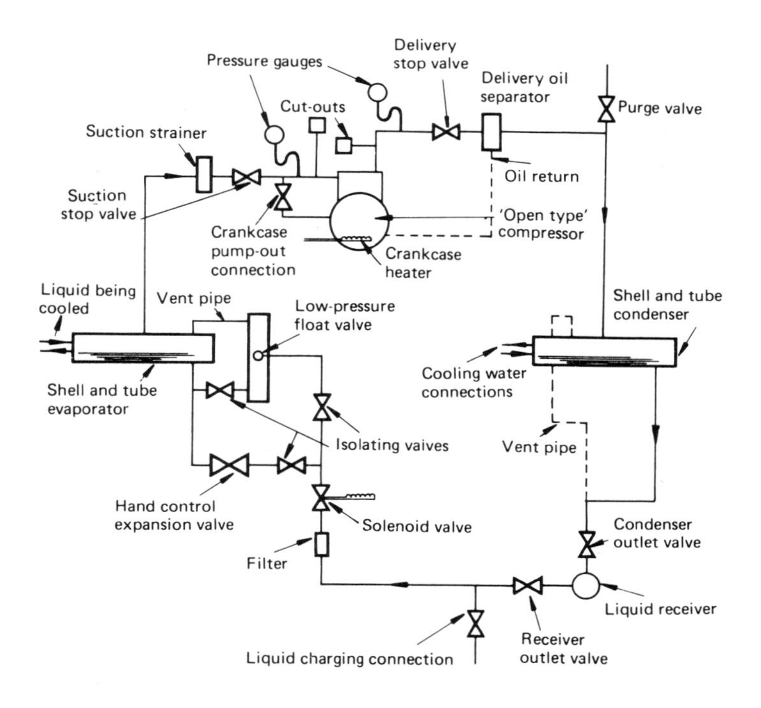

Figure 3. Layout of typical industrial refrigeration system

would normally be used as the refrigerant. The plant is being used for cooling brine which would then be circulated through pipe coils in the cold rooms or to ice making tanks.

Expansion Valve

The requirement with a plant of this type is for a valve that will maintain a constant liquid level in the evaporator, be capable of passing very large quantities of refrigerant and operate without superheat control.

The low pressure float valve shown meets all of these requirements. It does, however,

(a) Require periodic attention. A hand control expansion valve is therefore fitted in a by-pass system as shown. This stand-by valve is used to control the refrigerant flow while the main valve is being maintained, in order that normal operation of the plant can continue.
(b) Not close fully. When the compressor stops this valve frequently continues to pass refrigerant. This results in too high a liquid level in the evaporator and the possibility of unevaporated liquid being drawn into the compressor suction when it re-starts.

Solenoid Valve

This is fitted to the liquid supply pipe before the expansion valve to overcome the problem described in (b) above. Liquid passing into the evaporator during the compressor 'off cycle'.

It is normally wired in series with the compressor motor so that when the compressor stops the solenoid valve closes and no further liquid can flow into the evaporator. The action of this valve is fully described in SECTION F CONTROLS.

Suction Strainer

Industrial systems do not require the same degree of cleanliness as commercial. The valve and other openings are very much larger so that obstructions are less likely to occur.

Steel pipe is normally used, which corrodes and passes 'pipe scale' into the system. This is removed at the compressor, to prevent damage to its valves, by a wire strainer. The strainer, which is easily removed for cleaning, is held in a vertical steel cylinder. Any unevaporated liquid carried over from the evaporator is trapped in this cylinder to prevent it being drawn into the compressor. Arrangements are then provided to evaporate this liquid (frequently by passing the liquid pipe through it, as described in suction-liquid heat exchangers under commercial systems).

Compressor Fittings

Industrial compressors are normally fitted with the following attachments:

STOP VALVES — These are fitted at both suction and delivery to isolate the compressor. Unlike the service valves fitted to commercial compressors they cannot always be back seated.

CRANKCASE PUMP-OUT CONNECTION — This is a connection provided between the crankcase and the suction pipe fitted with an isolating valve that is normally closed.

When it is required to reduce the pressure in the crankcase, to carry out maintenance and repair work, the procedure is as follows.

Slowly close the suction stop valve and at the same time open the pump out line isolating valve. The compressor will now 'pump out' refrigerant vapour from the crankcase.

PRESSURE GAUGES These are permanently fitted to both the suction and delivery. They indicate the pressure and also, on a separate scale, the corresponding temperature (boiling point) for the refrigerant used in the system.

CRANKCASE HEATER When the compressor first starts any liquid refrigerant that has condensed in the crankcase will boil violently, due to the sudden reduction in pressure. This results in a foaming action, in which a mixture of oil and refrigerant vapour is carried past the piston rings and onto the top of the pistons. To prevent this occurring the crankcase oil is frequently heated during the time that the compressor is stopped. Only a small amount of heat, supplied by an electrical heating element, is necessary to avoid the problems of liquid condensing in the crankcase during the compressor 'off-cycle'.

Purge Valve

There is a greater tendency for 'non-condensable gases' to form in large industrial systems, particularly those using ammonia as the refrigerant. These gases raise the condensing pressure and must be removed periodically. Hand operated purge valves may be used but no larger plants a fully automatic purging arrangement is more reliable.

Liquid Cooling Evaporators

There are two different arrangements for cooling liquids:

(a) The liquid to be cooled is passed through a cylinder in which the refrigerant is evaporated INSIDE an arrangement of tubes. This is usually known as a 'dry evaporator' and refrigerant control is by thermostatic expansion valve.

(b) The liquid to be cooled is passed through a series of tubes fitted into the ends of the cylinder. The liquid refrigerant is evaporated on the OUTSIDE of the tubes and it is therefore known as a 'flooded evaporator'.

(c) This is the arrangement shown in Fig. 3 from which it can be seen that approximately half of the tubes are immersed in the liquid refrigerant. This gives a very high heat transfer rate and rapid cooling. Refrigerant control is usually by 'float valve' as shown.

Water Cooled Condensers

The high cooling capacity of an industrial plant requires a condenser that is capable of removing heat rapidly. Water cooled condensers have very high heat transfer rates and operate at a lower condensing pressure than air cooled types. They are therefore normally used for industrial work.

The construction is similar to the flooded evaporator. Cooling water is circulated through several passes of tubes with the refrigerant condensing on the outside. The end covers are removeable, so that the tubes can be cleaned and replaced.

Commercial Components

The following components, previously described under commercial refrigeration systems, are also usually fitted to the type of industrial system shown in Fig. 3.

LIQUID PIPE FILTERS (These do NOT incorporate a drying agent for ammonia plants)
HIGH AND LOW PRESSURE CUT-OUTS
DELIVERY OIL SEPARATORS
LIQUID RECEIVERS AND CHARGING CONNECTIONS

SECTION B

TOOLS AND ACCESSORIES FOR SERVICING AND INSTALLATION WORK

1 ENGINEERING HAND TOOLS

The following, general purpose, hand tools are used for servicing and installation work.

Spanners

(a) *Open-end ring type*. Sizes and dimensions (BSW, A/F, METRIC) to suit the equipment.
(b) *Adjustable*. For use only when the correct size of open-end is unavailable.
(c) *Self-gripping*. Referred to generally as the mole grip or wrench.
(d) *Socket set*. With ½ in square drive. Requirement as in (a) above.
(e) *Socket set*. With ⅜ in square drive (for smaller fastenings). Requirement as in (a) except that BA, A/F and METRIC are used.

Screwdrivers

A selection of Flat nose and Phillips types is required. Use only a screwdriver with the correct width of blade or size of head to fit the screw.

Select a screwdriver of a suitable length to apply the force required.

When working on electrical connections and controls ALWAYS use a screwdriver with a well insulated handle.

Pliers

For most work, 6 in pliers are satisfactory. The following types are required.

Side-cutting	Long nose	Electricians (well insulated)
Wire stripping	Pipe crimping	Circlip removal

Pliers are available that meet two, or more, of the above requirements.

Hammers

Ball Pein ½ and 1 lb (250 and 500 g) are most frequently used, rubber or plastic faced for use on easily damaged surfaces.

Files

Double cut files are used for most work and the following range, with well-fitting handles, is recommended.

Flat fine 6 in	Round fine 6 in	Flat medium 8 in (20 cm)
Round medium 8 in	Half round medium 8 in	

Hacksaws

(i) Adjustable metal frame complete with the following blades.

Teeth per inch (per cm)	*Materials*
18 (7)	Mild steel. Soft cast iron. Non-ferrous. Thickness ¼ in (6 mm) and above.
24 (10)	Alloy and high tensile steels. Also mild steel and non-ferrous $\frac{1}{8}$ to ¼ in (3 to 6 mm) thick.
32 (13)	Mild steel, non-ferrous, sheet metal below $\frac{1}{8}$ in (3 mm) thickness.

(ii) Junior or 'mini' type. For metal cutting in confined spaces.

Woodsaws

Tenon 14 in for general use.
Floor board 12 in for lifting T and G Boards.

Miscellaneous

ELECTRIC HAND DRILL	½ in (13 mm) chuck, suitable for local supply voltage and frequency.
SHEET METAL SNIPS	(a) Straight nose for plain cutting. (b) Curved nose for cutting radii.
ALLEN KEYS	For removing grub screws from fan blades, motor pulleys etc. (a) Imperial size $\frac{1}{16}$ to ¾ in. (b) Metric 1.5 to 22 mm.
POP RIVETER	For fastening work when one surface only is accessible.
CHISELS	Hardened steel, for metal cutting in ¼, ½ and ¾ in (6, 12, 20 mm) sizes.

PULLEY EXTRACTORS	(a) 3 leg type for pulleys up to 4 in (100 mm) diameter. (b) 2 leg type for pulleys between 4 in (100 mm) and 12 in (300 mm) diameter.
CENTRE PUNCH	4 in (100 mm) with plastic sleeve on handle.
WIRE BRUSH	8 in (200 mm) for general cleaning work.
BRISTLE BRUSH	8 in (200 mm) for condenser fin cleaning.
DRILLBITS	A range of sizes and dimensions (Imperial and/or Metric) to suit the equipment. High speed steel for metal cutting, masonry and wood types.
OIL CAN	Of the pump action type with an extended flexible nozzle.
KNIFE	5 in (130 mm) with replaceable blades.
EXTENSION LEAD	For electrical tools. 50 ft (15 m) minimum length.
INSPECTION LAMP	For mains supply with insulated handle.
BOLT EXTRACTORS	For the removal of broken bolts and studs. (Sizes to meet the equipment.)

2 MEASURING EQUIPMENT

The following represents the minimum requirement for measuring and testing in the field. Special purpose equipment, infrequently used on normal work is NOT included.

STEEL TAPE MEASURE	Flexible type 10 ft (3 m) long, graduated in inches and millimetres.
STEEL RULE	12 in long, graduated in inches and millimetres.
FEELER GAUGES	In Imperial and Metric sizes.
ENGINEERS SQUARE	8 in (200 mm) long, graduated in inches and millimetres.
CALIPERS	6 in (150 mm) long. Inside and outside measuring types.
MICROMETERS	Outside measuring type. Imperial range 0 to 1 in. Metric range 0 to 25 mm.
ENGINEERS LEVEL	12 in (300 mm) long, aluminium alloy construction.
POCKET TYPE THERMOMETER	6 in (150 mm) long in carrying case with pocket clip. Range – 20 to + 120 °F (–10 to 50 °C)
POCKET BALANCE	Suitable for weighing refrigerant cylinders, 50 lb (23 kg) capacity.
SLING PSYCHROMETER	Range 30 to 110 °F (–1 to 45 °C) with relative humidity conversion scale.
SYSTEM ANALYSER	For testing, evacuating and charging commercial plant. Fitted with low pressure (compound) and high pressure gauges, graduated as follows: 30 in (760 mm) vacuum to 100 lb/in^2 (7 bar); 0 to 500 lb/in^2 (18 bar). Temperature scales for the refrigerants in use and

fitted with restrictor stops and recalibrators. Colour coded to match charging lines (see Fig. 4).

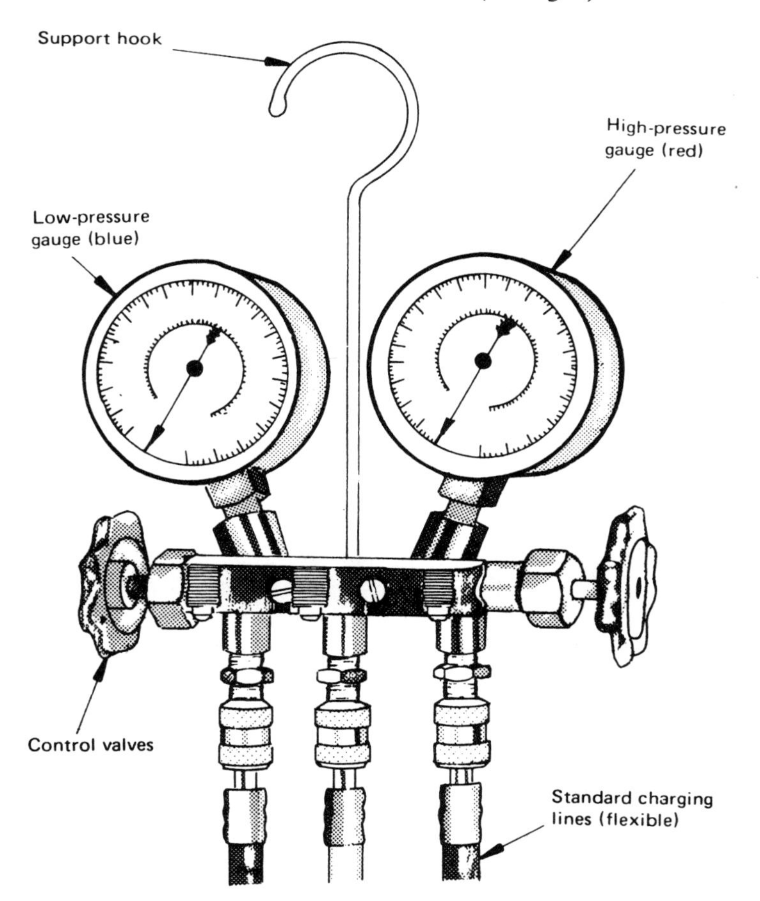

Figure 4. System Analyser

TEMPERATURE ANALYSER	Portable, battery operated, type. Range –50 to +150 °F (–50 to 65 °C). Suitable for, and complete with, at least 3 measuring probes.
ELECTRONIC LEAK DETECTOR	Portable, battery operated, suitable for all fluorocarbon vapours. With visual and audible leak indication.
VOLT WATTMETER	Fitted with separate scales, enabling wattage to be determined at a given voltage. Suitable ranges are 100 to 250 V AC, with wattage scales 0 to 500, 0 to 2500, 0 to 5000.
VOLT OHM METER	Battery operated 'clamp on' type for measuring resistance and continuity. A suitable range is 0 to 250 V AC, 0 to 5000 ohms.

3 SPECIALIST TOOLS AND ACCESSORIES

FLEXIBLE CHARGING LINES	Standard 3 ft (1 metre) hoses with knurled hand connections. Coloured, for ease of identification, red for high pressure, blue for low pressure, yellow for charging. Use for commercial work.
BENDING SPRINGS	Internal and external to suit the size of pipes in frequent use (See SECTION D PIPE BENDING).

PIPE (TUBE) CUTTERS

The following should be carried: (a) ¼ to 1 in (6 to 25 mm) diameter (b) $\frac{5}{8}$ to 1½ in (16 to 38 mm) diameter (c) MINIATURE, for cutting where access is difficult, $\frac{1}{8}$ to $\frac{5}{8}$ in (3 to 16 mm) diameter.

FLARING TOOL

Suitable for making SAE flares on soft copper pipe in the sizes described in SECTION D COPPER PIPE.

CAPILLARY TUBE CLEANER

Hand operated, hydraulic type suitable for clearing obstructions in capillary tubes (see Fig. 5).

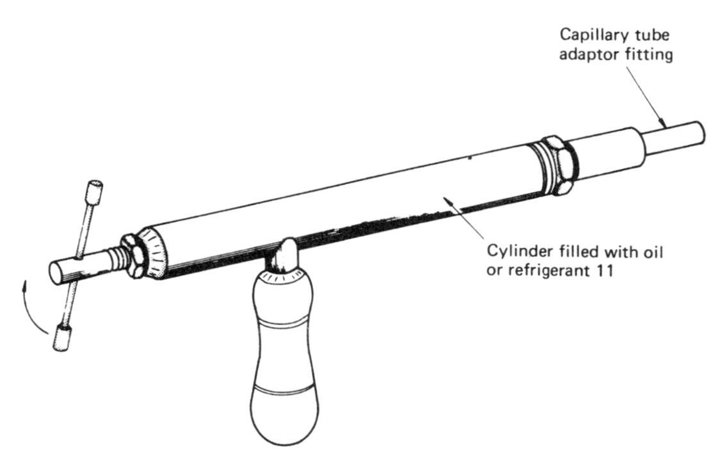

Figure 5. Capillary tube cleaner

RATCHET SPANNERS

For operating service valves. Reversible type ¼ in and $\frac{3}{8}$ in square drive.

FIN COMBS

For straightening and cleaning evaporator and condenser fins. A range of combs should be carried suitable for fin spacings between 4 and 16 to the inch (6 mm down to 1.5 mm between fins).

SOLDERING AND BRAZING EQUIPMENT

The oxygen-butane type is recommended for safe, accurate high quality work. It is provided complete with cylinder, pressure regulator, hoses and torch (with a range of nozzles to suit most work).

VACUUM PUMP

A lightweight portable pump, for system dehydration, complete with Torr gauge. Its capacity will be determined by the size of the refigerating system but it should be capable of drawing a vacuum of at least 1 Torr in not more than 30 min.

CHARGING CYLINDER

Usually marketed under the trade name DIAL-A-CHARGE, this enables a system to be charged quickly and accurately. It should be graduated for those refrigerants in use and incorporate a means of compensating for volume fluctuations with change in temperature.

Capacities are approximately 2½ and 5 lbs (1 and

2.5 kg) of liquid refrigerant. A model fitted with an electric heating element is preferable.

SCHRADER VALVE CORE REMOVER

For use when replacing a faulty valve or to improve the flow when evacuating the system (see Fig. 6).

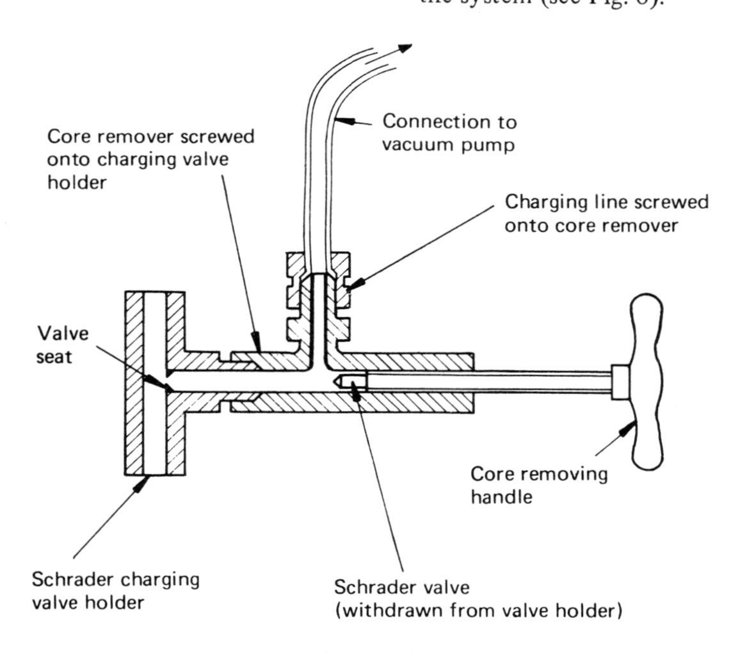

Figure 6. Schrader valve core remover

ELECTRIC TEST LAMP

Used to determine that there is a flow of current and that sockets are correctly wired.

JUMPER LEAD — Used to determine faulty electrical components by 'shorting across' or by-passing them.

4 SAFETY EQUIPMENT

PROTECTIVE GLOVES WELDING GLOVES
WELDING GOGGLES
SAFETY GOGGLES (For use when handling liquid refrigerant)
EAR PROTECTORS FIRE EXTINGUISHER FIRST AID KIT
RESPIRATOR-FACE MASK (Filter type for use when the air contains fumes or dust)
SELF-CONTAINED BREATHING APPARATUS (for use in industrial ammonia plants).

SECTION C

MATERIALS AND THEIR APPLICATION

1 SOLVENTS AND CLEANING

Contamination of the system must be avoided. This is achieved by thorough cleaning of all components and flushing the sytem.

Oil and Grease Removal (Field type methods)

(a) Spirit type cleaners
These operate on the solvent principle and the following are widely used:

KEROSENE (PARAFFIN)	Safe to work with but leaves a residue on the surface.
MINERAL SPIRIT (Proprietary types)	Has a higher fire hazard than kerosene but leaves a clean surface.
CARBON TETRACHLORIDE	Although a popular cleaning agent its use is NOT recommended. It is extremely TOXIC.

Other substances which are hazardous to use are PETROL, PROPANE, BUTANE and ALCOHOL.
Use a safe cleaner, in small quantities, with good ventilation and avoid open flames.

(b) Emulsion cleaners
These proprietary agents are neither toxic nor inflammable. They mix with oil and grease to form an emulsion which can be removed with water. Components cleaned in this way must be carefully dried (See SECTION H). A light film of oil may remain on the surface. This can, where necessary, be removed by a SMALL quantity of spirit cleaner.

System Flushing

Carried out mainly to remove acid contaminants following the 'burn out' of hermetic compressors. The entire system must be flushed either with the refrigerant used or, preferably, R11 which has a better scouring action.

R11 should also be used to flush either a system that has become contaminated in any other way, or a capillary tube after the removal of an obstruction. (See SECTION G).

ACETIC ACID (or proprietary acids) for cleaning water systems.

2 REFRIGERANTS

R11 A low pressure refrigerant with a high compressor displacement. Used for air conditioning work (with centrifugal compressors) also for flushing and cleaning (See Sub-section 1 above).

R12 Operates at moderate pressures. Mixes readily with oil which simplifies the problem of oil return to the compressor. Moisture is only slightly soluble in the liquid so that the systems must be kept thoroughly dry. Used for domestic and commerical work.

R13 A low temperature refrigerant for use at –40 °F (–40 °C) and below. The evaporating pressure, even at these low temperatures, is normally above atmospheric. Leakage inwards, resulting in system contamination, is therefore avoided. The refrigerant does not mix with oil and it is often difficult to arrange a satisfactory oil return. Used for low temperature work only.

R22 Compressor displacement is lower than when using R12, which reduces the compressor size. Operating pressures and compressor delivery temperature are higher and suction vapour superheat must be kept as low as possible. At low temperature the oil separates from the refrigerant. It is therefore essential to use delivery oil separators for low temperature work. Used as an alternative to R12, normally on larger plants.

R500 Used with compressors designed for 60 Hz but operating on 50 Hz supply. The increase in cooling capacity obtained, as compared with R12, compensates for the loss due to speed reduction. Used, for compressors described above, for domestic and commercial work instead of R12.

R502 Combines the best properties of R12 and R22. Moderate pressures with smaller compressors and a lower delivery temperature. Oil return to the compressor is more difficult than with R12 or 22. Used for frozen food applications.

R503 & 504 These are low temperature refrigerants that have better characteristics than R13. Oil return, in particular, is very much better and compressor sizes are smaller. Used for low temperature work, as a replacement for R13.

All of the above refrigerants, the FLUOROCARBONS, are completely safe. They are not toxic, poisonous, flammable or explosive. Avoid working in high concentrations as suffocation can occur. Do not allow large quantities to come into contact with high-temperature surfaces as refrigerant breakdown, and the formation of poisonous vapour, may occur. Make sure that good ventilation is provided when carrying out repair work.

R717 (AMMONIA). A dangerous refrigerant. It is poisonous, toxic, flammable and explosive. It should be handled carefully and contact with the skin avoided, especially the eyes.

The refrigerant is highly soluble in water, into which vapour should be purged, NOT to the outside air.

Used for large cold stores and ice making plant.

R744 (CARBON DIOXIDE). A completely safe refrigerant but operating pressures are extremely high and the compressor power is approximately twice as high as for most other refrigerants.

Used for marine work and in the frozen state, as dry ice, for refrigerated transportation.

R40 (METHYL CHLORIDE) and R764 (SULPHUR DIOXIDE). The refrigerants are dangerous; handle them carefully. Discharge the vapour into the outside air.

Sulphur dioxide is highly corrosive. Avoid contact with the skin, especially the eyes. Methyl chloride is flammable and has an anaesthetic effect.

Methyl chloride can be replaced by R12, without change of cooling capacity, provided that the system is thoroughly flushed after discharging the old refrigerant.

These refrigerants are no longer used for new work but may be met on older equipment.

3 LUBRICATING OILS

Use only refrigerant quality oils. They have been carefully prepared to meet the operating conditions of the system.

Use the type of oil recommended by the supplier of the equipment.

Oil is supplied 'moisture free'. It MUST be kept dry to prevent moisture from entering the system. Use small containers and keep them tightly sealed. Refrigerant oil will absorb moisture from the air. Ensure that a low temperature oil is used when operating at temperatures of –40 °F (–40 °C) and below, to avoid problems of wax deposits.

Use general purpose oils for normal lubrication of bearings, hinges, etc. in commercial work and penetrating oils for loosening rusted components.

4 SYNTHETIC REPAIR MATERIALS

These are widely used for the repair of leaks in commercial work when it is desirable to avoid high temperatures (as in soldering and brazing). There two types.

(a) *The two part adhesive*
Consists of an epoxy resin and a hardener. When mixed they harden, at room temperature, to give a permanent reliable joint. The material can be used either to repair a small hole or to fix a patch over a large one. To obtain satisfactory results the manufacturer's instructions MUST be strictly adhered to.
Use only those which are provided by suppliers of refrigeration materials. Other, similar, materials are frequently found to have poor adhesion, incompatibility with refrigerants, deteriorated due to extended storage and may be porous.

(b) *The one part adhesive*
This is supplied in stick form and applied using low temperature heat. It is suitable for use with all refrigerants and bonds well to most metals.
Used mainly for the repair of small holes or leaks especially when the system has to be taken straight back into service. There is no 'curing time' for the material to harden, unlike the two part material.

5 JOINTING MATERIALS

Gaskets

Soft materials used to prevent leakage between mating surfaces such as compressor head to valve plate, valve plate to compressor body, service valve to compressor etc.

Various materials are used. Compressed asbestos fibre and synthetic jointing is suitable for commercial work. Thicknesses available are $\frac{1}{64}$, $\frac{1}{32}$, $\frac{1}{16}$, $\frac{1}{8}$ in (0.4, 0.8, 1.6 and 3.2 mm). Larger industrial plants use cork sheet, available in thicknesses $\frac{1}{16}$, $\frac{3}{32}$ and $\frac{1}{8}$ in (1.6, 2.4 and 3.2 mm). Metal is sometimes used. Lead and aluminium are both soft and not susceptible to corrosion; these materials are frequently used on ammonia plants for pipe flange joints.

Pipe Joints

These sometimes have to be made with threaded screwed connections for industrial applications. Even when the thread has a taper form, which helps to prevent leakage, a jointing material applied to the thread is advisable. Suitable materials are

Jointing compound (Proprietary name, Hermetite). Easy to apply and suitable for water joints. Two types: Red remains soft; Green hardens to give a permanent joint.
Jointing tape (PTFE). Wound onto the male thread. Suitable for use with all fluids used in refrigeration work.

Pipe Supports

The type of saddle used for supporting refrigerant pipe is shown in Fig. 7.

6 DRYING MATERIALS

The following are in common use: ACTIVATED ALUMINA, SILICA GEL, MOLECULAR SIEVES.

The material (frequently termed a DESSICANT) is contained in a cylindrical shell, with a filter element at each end. It is located in the liquid pipe immediately before the expansion valve to remove all traces of moisture which would otherwise freeze at, and block, the valve.

Activated alumina sometimes disintegrates and allows a fine powder to enter the system.

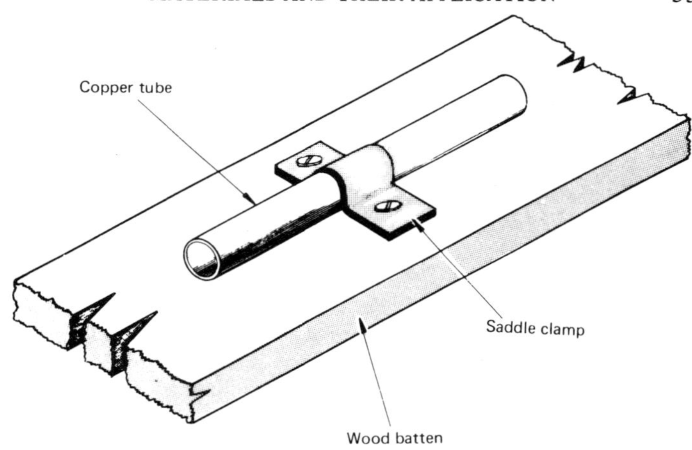

Figure 7. Refrigerant pipe support

Molecular sieves have a higher moisture retaining capacity than the other materials. Moisture in the system also causes corrosion. Refrigerants can absorb different quantities of moisture before a danger point is reached. For this reason, driers fitted to R22 and R500 plants should be approximately FIVE times the size of those fitted to R12 and R502.

With chemical driers they sometimes become warm as moisture is absorbed and in the absence of a sight glass, this is a useful indication that an obstruction is present at the drier.

If a drier becomes warm, due to shortage of refrigerant, or high temperature, it will frequently release moisture into the system.

A drier that blocks, causing a restriction, will become COLD.

Acid absorbing driers are used following the 'burn out' of an hermetic system. They normally contain silica gel with a molecular sieve core. These are placed in the compressor suction pipe and removed when all contaminants have been absorbed (See SECTION G).

Large industrial plants are normally fitted with a filter-drier whose filter and drying agent can be changed, rather than a throw-away component which is cheaper for smaller commercial plant.

7 PIPING MATERIALS

Most refrigeration work is carried out using metallic pipe (See SECTION D).

PLASTIC TUBE, usually polyethylene, is easy to work and is being increasingly used. Although mainly restricted to water systems fittings are available for connection to refrigeration components. It is unsuitable for either very low or high temperature work. Safe limits are approximately −100 to +200 °F, −73 to +93 °C.

Normally available to the following sizes $\frac{1}{8}$, $\frac{3}{16}$, $\frac{1}{4}$, $\frac{5}{16}$, $\frac{3}{8}$, $\frac{1}{2}$ in (3, 5, 6, 8, 10, 13 mm).

FLEXIBLE TUBE is used particularly on mobile work for liquid and compressor suction pipes.

It is constructed from materials that do not ‘age harden’ but remain flexible and have a very low leakage rate.

A range of fitting is available for the connection of flexible hose to standard flare type fittings.

8 INSULATING MATERIALS

The important properties and normal usage of insulating materials in common use are as follows.

Expanded polystyrene/polyurethane. Light, easy to erect and can be expanded *in situ*.)

Resists moisture penetration but is flammable, so take care when soldering or brazing. Low density types have higher resistant to heat flow but are more easily damaged (protection required). Used for most insulating work where high loads are not involved (See CORK) or there is no fire risk.

Cork. Compressed slabs are strong, for insulating type materials.

Absorbs moisture and therefore requires waterproofing, flammable (soldering/brazing risks). Used for large cold store floor insulation (particularly where fork lift trucks are used). It is able to withstand high floor loadings without deforming.

Laminated materials. Layers of material (usually aluminium foil) with air spaces between. These materials are light, non-flammable and do not absorb moisture. They are easily damaged, by crushing, and must be protected on each side. Used for refrigerated containers and transportation work.

Mineral fibres. Cheap and non-flammable but absorb moisture and are easily damaged. Used for low quality work and older installations, but being replaced by polystyrene/polyurethane.

Expanded rubber. Supplied in sections, circular, or to fit a component. Flammable but does not absorb moisture.

A proprietary material 'Armaflex Plus One' is fire-resistant and meets the requirements of most fire regulations. Used for insulation of pipes, valves and other fittings.

SECTION D

PIPING AND JOINTING WORK

1 COPPER PIPE

Used for all systems where the refrigerant is a synthetic fluoro-carbon and in most commercial work.

Referred to by its outside diameter (OD). Copper pipe used for plumbing, and similar, is referred to by inside diameter or in millimetres. This must not be used for refrigeration work it will contaminate the system.

Refrigerant quality pipes

(a) *Soft annealed*

Easy to bend by hand and used for small installations.

Flare type connections are used (see Pipe Joints, below).

Available in the following sizes: $\frac{1}{8}$, $\frac{1}{4}$, $\frac{3}{8}$, $\frac{1}{2}$, $\frac{5}{8}$, $\frac{3}{4}$, $\frac{7}{8}$ in (3, 6, 10, 13, 16, 19 22 mm).

(b) *Hard drawn*

Strong and rigid. Suitable for larger diameters and long pipe runs.

It must NOT be bent. Use fittings for changes in direction.

Silver solder or brazing rod used for all connections.

Available in the following sizes $\frac{7}{8}$, $1\frac{1}{8}$, $1\frac{3}{8}$, $1\frac{5}{8}$, $2\frac{1}{8}$, $2\frac{5}{8}$, $3\frac{1}{8}$ in (22, 28, 35, 41, 54, 67, 80 mm).

2 STEEL PIPE

Used for industrial work and particularly where ammonia is the

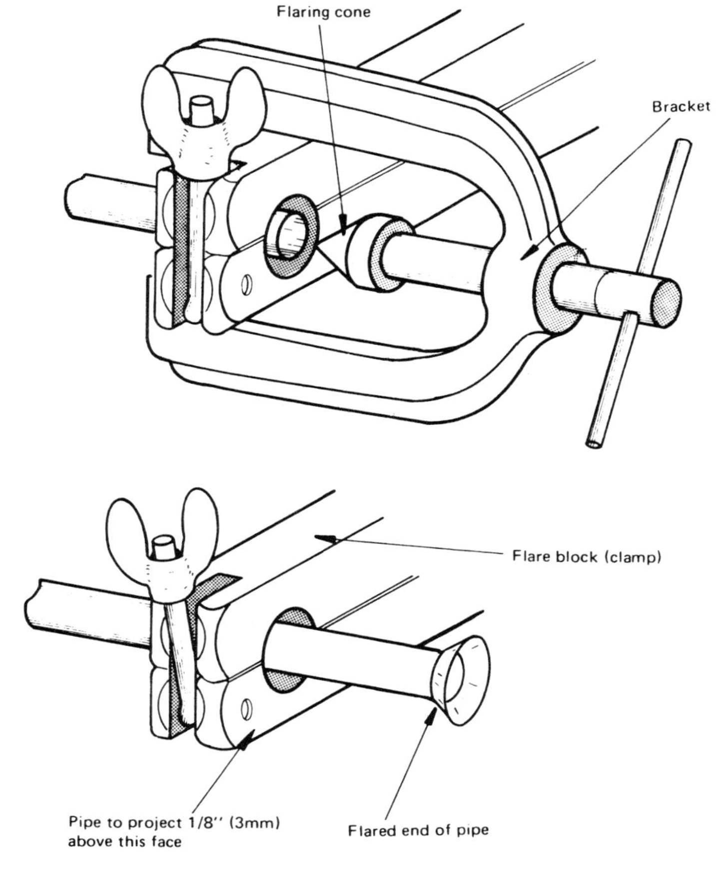

Figure 8. Flaring copper pipe

refrigerant as ammonia is highly corrosive to copper.

Connections are by welding except that components which have to be removed are connected by flanged joints.

Available in the following sizes $\frac{1}{8}$, $\frac{1}{4}$, $\frac{3}{8}$, $\frac{1}{2}$, $\frac{5}{8}$, $\frac{3}{4}$, 1, $1\frac{1}{4}$, $1\frac{1}{2}$, 2, $2\frac{1}{2}$, 3 in (3, 6, 10, 13, 16, 19, 25, 32, 38, 50, 64, 76 mm).

3 PIPE JOINTS

Flare Connections

Used for connecting soft copper to brass fittings (bends, tee pieces etc.)

Correct procedure for flaring copper pipe:

Cut to length with a wheel type cutter. Do not use a hacksaw.

Remove burrs with the pipe cutter blade. Do not allow copper cuttings to enter the pipe.

Place the flare nut on the pipe. It cannot be fitted after the end has been flared.

Fit the pipe into the clamp of the flaring tool. Allow it to project $\frac{1}{8}$ in (3 mm) above the face of the clamp (see Fig. 8).

Lightly oil the cone of the tool and then tighten it into the end of the tube.

Remove the pipe and check that the flare has been properly formed. Make sure there are no splits, indentations or wrinkles on the flared surface.

See that the end of the pipe is clean, dry and fits well before connecting it to the fitting.

Soldered and Brazed Connections

Use good quality silver solder only. Soft solder, as used in plumbing work, is NOT suitable. Brazing, using a brass filler rod, should be used to give the additional strength required with large pipes ($\frac{7}{8}$ or 22 mm and above).

Correct procedure for either soldering or brazing:

(a) *Preparation*

Clean the outside of the pipe and the inside of the fitting thoroughly with wire wool. Do NOT use emery cloth or other abrasive materials.

Check the fit of the pipe in the fitting. It must enter easily, without binding, but not be loose.

(b) *Purging*

Prevent the formation of scale on the inside of the pipe when heating by passing nitrogen through it.

Connect the pipe-work to a nitrogen gas cylinder through a pressure reducing valve. Allow sufficient nitrogen to flow through the pipe-work to displace the air inside through the other, open, end.

Check the flow of nitrogen before applying heat. A large flow is not necessary. Provided the flow can be felt on the hand, it is sufficient.

(c) *Applying Flux*

A flux must be applied to the surfaces to be joined to prevent the formation of oxides and assist the flow of solder. Use only a flux of the correct type, suitable for the materials being joined and the filler rod.

A thin, even, film of flux is necessary. Too little will give a poor bond; too much will contaminate the system.

Insert the pipe into the fitting for about $\frac{1}{4}$ in (6 mm). Apply the flux, with a brush, to the outside of the pipe. Push the pipe into the fitting, until it reaches the internal shoulder, then rotate it a few times to spread the flux evenly.

(d) *Supporting*

When all of the joints have been prepared the pipe-work assembly MUST be securely supported, where necessary, to prevent any movement during heating.

If any movement of a soldered/brazed connection occurs before the jointing material has cooled sufficiently to completely solidify a leaking joint will almost always occur.

(e) *Applying Heat*

Heat the joint evenly. Apply the flame to the pipe and fitting alternately. The fitting will need more heat than the pipe as it has a greater mass.

Do not allow the flame to touch the seam where the solder is to be applied. It will burn and destroy the action of the flux.

Do not overheat the joint. This will decompose the flux and affect the solder. The correct temperature has been reached when the solder melts. This can be easily found by touching the solder onto the hot metal from time to time while heating.

(f) *Aplying Solder*
When the pipe and fitting are both hot enough to melt the solder, touch the joint with the solder at several places.
Capillary action will draw the molten solder into the joint and around the pipe. When a complete ring of solder appears, all around the pipe, a sound leak-proof joint has been made.
Do not move the joint, or cool it rapidly, until the solder has completely solidified.

4 PIPE BENDING

This is carried out to give a change of direction without the use of fittings such as bends and elbows. It is quicker, cheaper and gives a larger radius which gives a better refrigerant flow.

Steel pipes are bent with special bending machines. These produce accurate bends without risk of damaging the pipe. Follow the supplier's instructions carefully.

Small diameter copper pipe is bent by hand using bending springs. These are used to prevent the thin wall of the copper pipe from kinking or flattening. Internal springs fit inside the pipe. They are particularly useful for making bends near to the end of the pipe, or even after the pipe has been flared. External springs fit over the outside of the pipe. This type is useful for making a bend in the centre of a long length of pipe. Pipe springs are available in the following sizes:

$\frac{1}{4}$ (external type only) $\frac{3}{8}$, $\frac{1}{2}$, $\frac{5}{8}$, $\frac{3}{4}$, $\frac{7}{8}$ in (6, 10, 13, 16, 19, 22 mm).

Useful Hints on Pipe Bending

(a) Ensure that the pipe is bent so that it does not place any strain on the fittings after it is installed.

(b) Do not attempt to bend a pipe to an inside radius less than FIVE times the diameter of the pipe.

(c) Make sure that the pipe is kept round for the entire length of the bend as kinks on the inside of the pipe restrict the flow of refrigerant.

(d) Bend the pipe slowly, carefully and gradually. Do not attempt to complete the bend in one operation as this increases the risk of the pipe breaking or distorting.

(e) When using bending springs bend the pipe initially rather more than required, then bend it carefully back to the required position. This makes removal of the bending spring easier.

(f) A small quantity of refrigerant oil applied to the surface of a bending spring will prevent it from rusting and make removal from the pipe easier. Springs can be removed more easily from the pipe after it has been bent if they are twisted during removal.

(g) Remember to remove external springs before flaring or jointing.

Pipe Installation Work

See SECTION H INSTALLATION TECHNIQUES.

SECTION E

ELECTRICAL INSTALLATION WORK

1 SAFETY PRECAUTIONS

Ensure that the electrical circuit is disconnected before starting work

Test the component. Make sure that it is not 'LIVE' and cannot be re-connected by another person.

Use insulated tools and rubber-soled footwear.

Try to avoid working in wet or damp locations.

Discharge capacitors (by short-circuiting) before handling them.

Make sure all electrical components are properly earthed.

Use electrical equipment at its rated voltage only.

Keep live electrical contacts and terminals covered by an insulated casing.

Handle old brittle electric cable carefully and replace it where possible.

Make sure that all electrical connections are correctly and securely fitted.

See that portable electric equipment is correctly connected to a 3-pin (earthed) plug and socket. Use only correctly rated fuses; the use of odd pieces of wire or other objects is dangerous.

Ensure that correct colour coding of all cables and wires is adhered to. The following table summarises the coding used in the UK for most of the wiring.

FUNCTION	COLOUR		
	Fixed cable	*Flexible leads*	
Earthing	green-and-yellow or green	green-and-yellow	(green*)
Neutral	black	blue	(black*)
Live (single phase)	red	brown	(red*)
Phase R of 3-phase	red	brown**	(red*)
Phase Y of 3-phase	yellow	brown**	(white*)
Phase B of 3-phase	blue	brown**	(blue*)

* These colours are out of date, but may be found in old equipment.
** Phase rotation marking, if required shall be by the application of numbered or lettered (not coloured) sleeves.

2 ELECTRIC SHOCK

Should a person receive a severe electric shock:

(a) If he is still in contact with the supply, either SWITCH OFF (if an isolator is convenient) or remove him with a non-conducting material (a chair, wooden stick or coat).
(b) If he is unconscious, apply artificial respiration. Personnel working with electricity should become proficient in an approved method.
(c) Send for professional medical attention.
(d) Keep the person warm and quiet. Cover burns to keep out air and dirt but do NOT apply any ointment or similar.

The above action must be carried out in the order shown as quickly as possible.

3 WIRING WORK

In refrigeration practice this is carried out in either PVC insulated cable, enclosed in a protective conduit, or with mineral-insulated cable.

(a) *Conduit work*

This is cheaper and usually preferred.

Fix the conduit securely using spacer-bar saddles (Fig. 9).

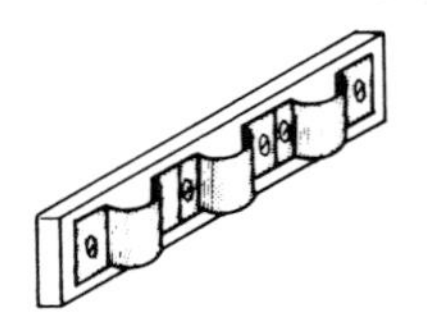
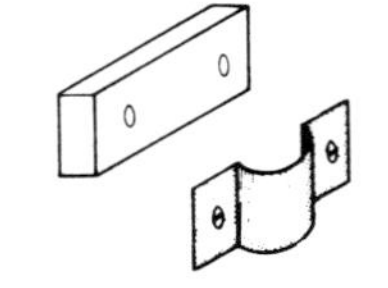
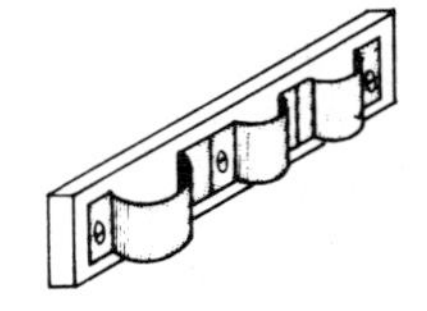

Figure 9. Conduit pipe supports

This makes the connection into boxes easier and prevents accumulation of dirt and moisture. Mark out the conduit runs carefully with a chalk line, avoid obstructions and possible damage later, keep all runs vertical or horizontal.

Allow for condensation inside by avoiding traps, fitting drains or arranging the conduit system to be self-ventilating.

If the conduit system is to act as an earth-continuity conductor ensure that all connections are electrically and mechanically continuous by fitting earthing clamps.

Plastic conduits are easier to join and bend and also more resistant to corrosion. They are, however, more easily damaged, become brittle when cold and soften under warm conditions.

(b) *Mineral-insulated cables*

These may be either copper or aluminium covered.

Aluminium covered is cheaper, lighter and will withstand higher temperatures. It is larger in diameter and more easily damaged (NOT recommended for surface use).

Copper (MICC) is normally preferred and a wider range of fittings is more readily available. DO NOT use copper fittings for aluminium cable or vice versa. The cable is supplied in coils and has been annealed. After running the cable from the coil pass it through roller-straighteners to remove all bends and kinks. Bending is by hand, although bending levers are available for large cables. DO NOT bend to a radius less than six times the outside diameter of the cable and soften it with a blow lamp if necessary.

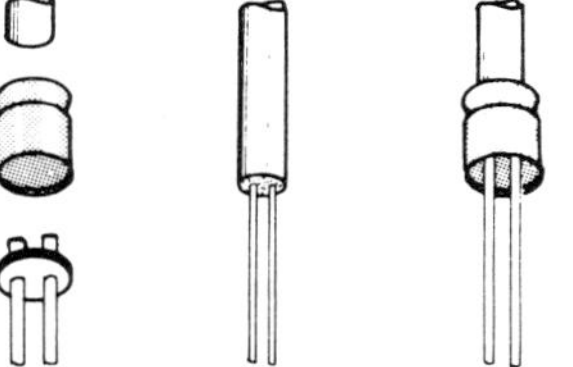

Figure 10. Mineral insulated cable seals

(c) *Terminating*

All cable terminations must be sealed, see Fig. 10, as the cable's insulating compound absorbs moisture. The procedure is:

(i) Prepare the cable end. Cut to length, allowing the required length of conductors for connecting up.

Mark the position where the outer sheath is to end and remove the excess.

Clean the conductors and remove all insulating com-compound.

(ii) Screw on the end pot using a pipe grip or pot-wrench tool. The pot has self-cutting threads. Push it squarely onto the cable and engage the threads by hand. Use a tool as the thread tightens and screw the pot down the cable until the internal shoulder reaches the cable end or the threads start to bind.

(iii) Fit the insulating sleeve and disc and apply the sealing compound. See Fig. 11. Press the compound firmly into

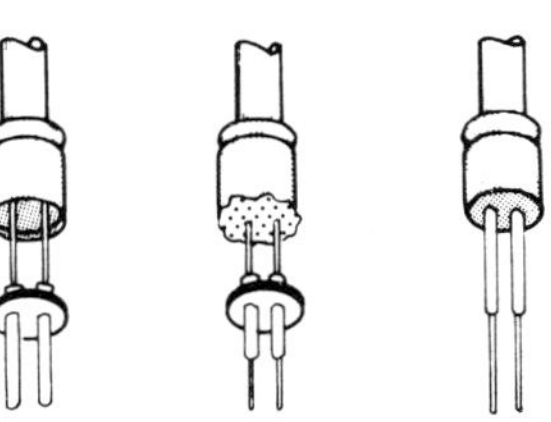

Figure 11. Application of the sealing compound

the pot, from one side only to prevent air pockets, and over-fill so that a small mound is formed outside the pot.

(iv) Crimp the end to effect a seal using a standard crimping tool. Wipe away any compound forced out of the pot and check that the disc has seated properly.

(v) Carry out an insulation test. If the seal is effective, an infinity reading should be obtained.

(vi) Position the gland to house the pot seal, connect and tighten up. See Fig. 12.

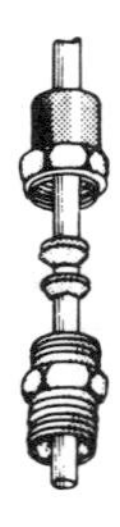

Figure 12. A completed cable seal

(d) *Installing cables*

Support, as necessary, with copper clips and saddles.

If a neat appearance is required the outside of the cable can be 'dressed' with a wood block and hammer. Be careful not to dent the cable.

Protect this type of cable from possible damage where it passes through a floor carry it in steel trunking or similar.

Protect it from corrosion (floors that are washed). When used underground, enclose it in a PVC sheath.

When connecting to a machine, or any unit subject to vibration or movement, make a loop or bend near to the termination.

When bending a cable to enter a gland into a fitting form an offset, allowing about 2 in (50 mm) of straight cable between the gland and bend. Withdrawal of the gland and nut from the fitting is then easier.

4 DISTRIBUTION BOARDS (for commercial work)

Standard capacities for single phase supply are 20, 30, 60, 100 and 200 A. Some patterns can be obtained in 45, 80 and 300 A capacity.

They are normally available with 2, 3, 4, 6, 8, 10 and 12 ways, although 24-way boards are available.

Use a board of adequate capacity with spare ways. Allow for possible alterations and extensions.

Select a distribution board that gives good access, with the cover removed, and with cable entry and earthing connection to suit the wiring system.

Terminate the wiring to suit the system and ensure that the cable is well protected.

Where possible, position the distribution board near to the centre of the load, where it is accessible and dry.

Select suitable fuses that will operate at not more than 1.5 times the current rating of the circuit.

Provide, on the inside cover of the distribution board, a clear indication of the circuit protected by each fuse and the rating for the circuit.

5 ELECTRICAL TESTING

Although it is essential to carry out the following tests on completion, it is often advisable to test at intervals during progress of the work. This is particularly the case on large installations where faults are easier to find and rectify before completion of all the work.

Verification of polarity

This test must be carried out to ensure that, in a two wire installation, all fuses and single-pole control devices are connected to the live conductor.

Earth continuity conductor test

The impedance or resistance of the earth continuity conductor must be tested to ensure it does not exceed 1 ohm.

Effectiveness of earthing

This test must be carried out to ensure that the earthing arrangements would result in satisfactory operation if a fault occurred.

Insulation resistance test

The insulation resistance to earth of the completed installation must be tested to ensure that it is not less than 1 megohm.

Ring circuit continuity

A test must be made to verify the continuity of all conductors, including the earth-continuity conductor, of any ring main circuit.

Testing methods

The above tests must be carried out in the order shown and following an approved method as laid down in 'THE REGULATIONS FOR THE ELECTRICAL EQUIPMENT OF BUILDINGS' issued by the Institution of Electrical Engineers (or, outside the United Kingdom, the appropriate local regulations).

6 CIRCUIT DIAGRAMS

When installing a new plant, or tracing an electrical fault on an

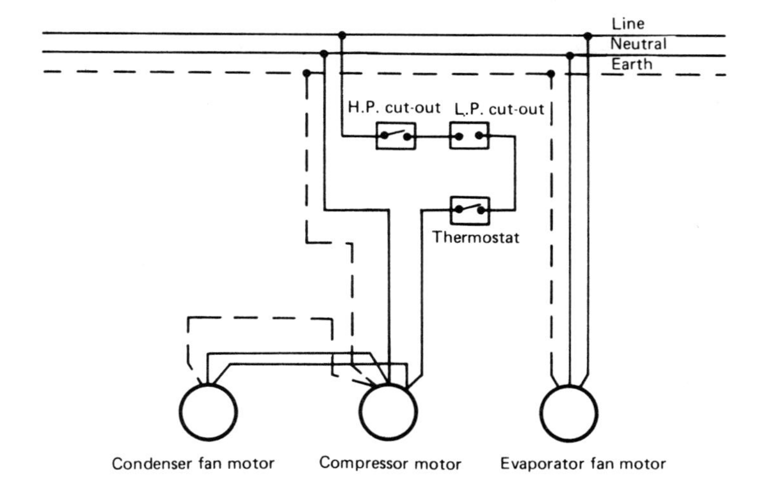

Figure 13. Single-phase wiring diagram. Evaporator fan running continuously. Condenser fan motor interlocked with compressor motor

existing one, always try to obtain the electrical circuit diagram. It will make the job easier, quicker and you are less likely to make mistakes.

If a circuit diagram is not available, draw one. Identify the electrical connections to the various components and prepare a logical sketch of the cable runs that you consider would be required for the correct, safe operation of the plant. If you are uncertain as to its accuracy have it checked by a competent engineer BEFORE starting work.

The following diagrams (Figs 13 and 14) will be found helpful, particularly since they are more clearly labelled than normal electrical circuit diagrams.

Commercial systems may be single-phase (up to about 1 HP or 1 kW) or three-phase. Industrial systems are always connected to a three-phase supply.

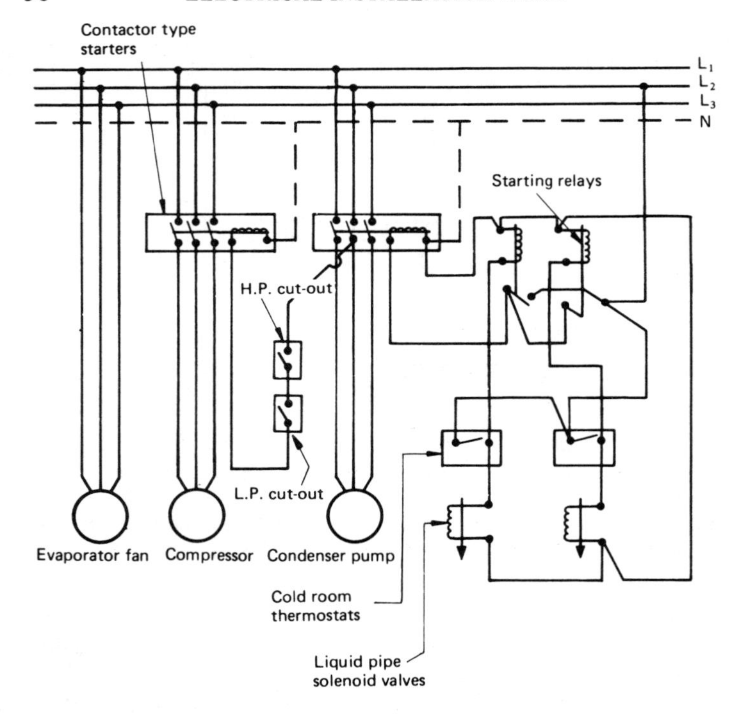

Figure 14. Three-phase wiring diagram for two cold rooms maintained at different temperatures

SECTION F

CONTROLS

The following controls are dealt with in this section.

Type of control	*Function of control*	*Element controlled*
Expansion Valves (All types)	Regulate flow of refrigerant	Evaporator
Thermostat	Start and stop plant	Compressor Motor Condenser Fan Motor Solenoid Valve
High Pressure Cut-out	Protect against overload	Compressor Motor
Low Pressure Cut-out	Protect against low temperature	Evaporator
Solenoid Valve	Permit or prevent refrigerant flow	Liquid line
	Permit or prevent water flow	Condenser
Water Regulating Valve	Control flow of water	Condenser

1 EXPANSION VALVES

Thermostatic

The most common, used for commercial and industrial work. It is the ONLY valve which, if properly adjusted, ensures that liquid cannot return to and damage the compressor.

(a) *Installation*
Position the phial on a vertical pipe with the groove fitted

against the side (on a horizontal pipe attach to the top at 10 or 2 o'clock).

Clean the outside of the pipe and remove all grease and moisture.

Do not position the phial near to fittings, or at any point where liquid can accumulate.

Fasten the phial securely to the suction pipe (using the clip provided).

Insulate the phial and suction pipe with rubber foam (see Fig. 15).

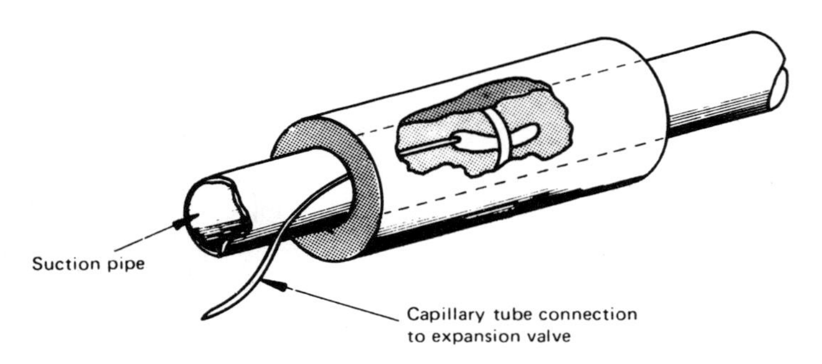

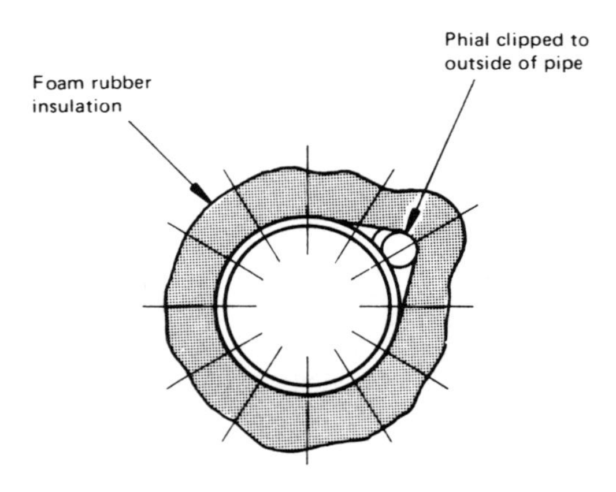

Figure 15. Installation of a thermal phial

(b) *Faults*

(i) Passing too much liquid — *Symptoms.* Excessive sweating or frosting of suction pipe. Evaporating pressure higher than normal. Compressor cylinder head cooler than normal.

Cause	*Remedy*
Valve out of adjustment	Re-adjust, slowly, closing the valve (evaporating pressure will reduce if adjustment is correct)
Phial not secured properly (or being affected by warmer air)	Examine for tightness, ice under phial and good insulation (See *installation* above) Re-fit to pipe correctly.
Wax deposits or ice crystals preventing valve from closing	Warm valve to melt ice (Pressure will immediately fall). Replace drier. Remove and flush valve, with R11, to dissolve wax. Ensure oil is suitable for evaporating temperature.
Valve needle stuck in the open position	Large valves, return to workshop for overhaul. Small valves, discard.

(ii) Not passing sufficient liquid — *Symptoms.* Evaporating pressure lower than normal. Compressor cylinder head warmer than normal. Loss of cooling effect.

Cause	*Remedy*
Valve out of adjustment	Re-adjust, in opposite direction to instruction above. Evaporating pressure should rise.
Valve inlet filter dirty	Remove and clean or replace.

Wax deposits or ice crystals preventing the valve from opening.	As in (i) above.
Reversal of phial charge	Warm head of valve carefully to return charge into phial.
Loss of phial charge	Large valves, replace power element. Small valves, discard.
Valve needle stuck in the closed position	Large valves, return to workshop for overhaul. Small valves, discard.

Constant pressure or automatic

Used mainly for commercial liquid chilling work, when the cooling load is fairly constant and low evaporating pressures must be avoided.

(a) *Faults*

(i) Passing too much liquid — *Symptoms.* See *Valve faults* above.

Cause	*Remedy*
Valve out of adjustment	Re-adjust and reduce flow to match load.
Diaphragm or bellows punctured.	Re-new of possible, otherwise replace valve.
Valve stuck in the open position.	Large valves, remove and flush with R11. Small valves, discard.

(ii) Not passing sufficient liquid — *Symptoms.* See *Valve faults* above.

Cause	*Remedy*
Valve out of adjustment	Readjust and increase flow to match load.
Inlet filter dirty	Remove and clean or replace.
Valve stuck in the closed position	See *Valve faults* above.

(iii) Erratic valve operation	*Symptoms.* Fluctuating evaporating pressure. Alternate flooding and starving of evaporator.
Cause	*Remedy*
Stabilising spring weak or broken	Large valves, overhaul and replace. Small valves, discard.
Push rods not moving freely	See above. Valve stuck in the open position
Valve too large	Compare valve and evaporator capacities. Fit valve of correct capacity.

Capillary tube

Fitted to domestic refrigerators, deep freezers etc.

For symptoms of blockage	See SECTION K
For method of clearing	See SECTION G

Low pressure float valve

Used with flooded evaporators to maintain a constant liquid level (industrial applications).

(a) *Faults*

(i) Passing too much liquid	*Symptoms.* Unevaporated liquid returns to the compressor causing sweating, frosting or knocking. Evaporating pressure higher than normal.
Cause	*Remedy*
Valve not seating properly	Remove and clean valve, seating and inlet filter. Remove and replace.

	Ball float punctured	Handle carefully, high pressure liquid may be trapped inside float.
	Operating linkage jammed	Dismantle, clean and lubricate.
	Valve passing liquid during compressor off-cycle	Fit solenoid valve, wired in series with compressor, in the liquid pipe before the valve. (Symptoms will occur on first start up only)
(ii)	Not passing sufficient liquid	*Symptoms.* Evaporating pressure lower than normal. Loss of cooling effect.

Cause	*Remedy*
Inlet filter dirty	Remove and clean.
Operating linkage not moving freely.	Dismantle, clean and lubricate
Float chamber 'gas locked'	Check vent pipe and remove obstructions. Insulate float chamber and pipe-work to maintain a constant temperature and pressure.

High pressure float valve

This valve is used for industrial applications and will also maintain a constant liquid level, provided that the plant is correctly charged. To operate satisfactorily the charge MUST be correct.

(a) *Faults*

(i) Passing too much liquid — *Symptoms.* See *Low Pressure Float Valves* above.

Cause	*Remedy*
System overcharged	Slowly purge refrigerant from the system until the correct operating conditions are obtained.

(ii) Not passing sufficient liquid	*Symptoms.* See *Low Pressure Float Valves* above.
Cause	*Remedy*
Inlet filter dirty	Remove and clean.
Float punctured	Remove and replace. See *Low Pressure Float Valves* above.
Operating linkage not moving freely	Dismantle, clean and lubricate.
System undercharged	Trace leak and repair. Add refrigerant CAREFULLY. Do not overcharge.
Float chamber 'gas locked'	See *Low Pressure Float Valves* above.
(iii) Complete loss of cooling	*Symptoms.* Evaporating and condensing pressures equalise.
Cause	*Remedy*
Valve stuck open due to dirt at the seat or linkage jammed	Dismantle, clean and lubricate.

2 THERMOSTATS

Wiring connections are shown in Fig. 13 SECTION E.

Method of Control

Space control
The sensing element is located in the air, or liquid, to be cooled. This gives closer control of cold storage or liquid chilling temperature.

Evaporator control
The sensing element is clamped to the evaporator and controls the evaporating temperature, which in turn controls the product temperature. Although minor fluctuations in storage temperature occur, this method is usually preferred (for applications above freezing) in order that off-cycle defrosting can be more easily accomplished.

Sensing Elements

Pressure bellows
A fluid filled tube or phial is attached to a bellows or diaphragm. Increase in temperature, at the point of control, increases the pressure of the fluid which acts through the bellows and a system of levers to close a pair of electrical contacts. Decrease in temperature has the opposite effect.
Bi-metal strip
The sensing element is in the form of a compound bar made from two different metals. One metal has a very high, the other a very low, coefficient of expansion. With changes in temperature, a movement will occur of the bar which is used to open and close electrical contacts, or actuate other mechanisms.

Faults

(a) Control element not securely attached to evaporator (an excessive temperature difference is necessary before the thermostat operates)
The element must be firmly clamped to the evaporator surface, which in turn must be clean and dry to give a good thermal contact.
(b) Control element has lost its charge (pressure bellows type)
To check, compress the bellows by hand. When fully charged, the internal pressure is approximately 100 lb/in^2 (6.9 bar). If the bellows move under finger pressure, the charge, or some of it, has been lost. Replace the bellows assembly, if accessible, otherwise change the thermostat.
(c) Worn, pitted or corroded contact points
These result in poor electrical contact. Check the thermostat circuit with a test lamp or ohmmeter. In an emergency contacts can be cleaned with a small file but worn or pitted contacts should be replaced.
(d) Poor electrical connections
For satisfactory operation these connections must be clean and tight. Check for continuity, as in (c) above.

Testing

See *Low Pressure Cut Outs* below.

3 PRESSURE CONTROLS

The operating principle is the same as the pressure bellows thermostat except that the bellows is actuated by refrigerant vapour direct from the evaporator. They are fitted to the high pressure side of the system (4) and frequently to the low pressure side also (5).

4 HIGH PRESSURE CUT OUTS

These are safety devices fitted to the compressor discharge and wired in series with the compressor motor to avoid excessive pressure. They must always be fitted where water cooled condensers are used, to avoid excessive pressure if the water supply fails, and the connection to the cut out must be taken from a point in the discharge where it cannot be inadvertently isolated (for example, by closing the compressor delivery stop valve).

Pressure Settings

The control must be set to cut-out (shut down the compressor) at a pressure well above the maximum condensing pressure on full load, but well within the safe operating pressure of the high pressure side of the plant. In temperate climates, the highest condensing temperatures are unlikely to exceed 120 °F (50 °C) (AIR COOLED CONDENSER), 100 °F (40 °C) (WATER COOLED CONDENSER).

Acceptable cut-out settings are, therefore,
200 lb/in^2 g (138 °F, 59 °C) for an air cooled R12 plant.
275 lb/in^2 g (121 °F, 49.4 °C) for a water cooled ammonia plant.

The difference in pressure between cut-out and cut-in (the differential) will depend upon the application. The following must be considered.

SMALL DIFFERENTIALS	Result in compressor short-cycling with attendant damage to electrical equipment.
LARGE DIFFERENTIALS	Result in excessively long off-cycles which may result in a rise of storage temperature and deterioration of the product.

Faults

(a) *Electrical.* Generally as for THERMOSTATS (See paragraph 2 above)
(b) *Mechanical.* Failure of pressure bellows, obstruction of connecting pipe, loose, worn or damaged linkages (see paragraph 5 (b) below).

Testing

See paragraph 5 (c) LOW PRESSURE CUT OUTS below.

5 LOW PRESSURE CUT OUTS

These are identical to high pressure cut outs except that they break the circuit to the compressor motor on a fall of evaporating pressure. Low evaporating pressures result in low temperature with possibility of damage to the product and plant (liquid chilling applications), and create a possibility of system contamination if the pressure falls below atmospheric.

(a) **Pressure Settings**

The control must be set to cut out at a pressure well below the lowest normal evaporating pressure on minimum load, but above a pressure where any of the following are likely to occur.

Freezing, with subsequent damage, of any liquid being cooled.
Damage, caused by low temperature, to the product in cold storage.
Blockage, by ice and frost, of chill room evaporators.

Small differentials must be avoided, as with high pressure cut outs, to prevent short-cycling damage. Large differentials will also result in the problems of long off-cycles described under high pressure cut outs.

(b) **Faults**

(i) *Electrical.* Generally as for thermostats (See paragraph 2 above).

(ii) *Mechanical.*

FAILURE OF PRESSURE BELLOWS.	If suspected test for leaks and replace bellows (large control). Normally replace control.
OBSTRUCTION OF CONNECTING PIPE.	The small pipe connecting the evaporator to the control occasionally gets partially blocked with pipe scale and sludge. This results in erratic operation of the control. To remedy, isolate and remove pipe, then flush with R11.
LINKAGES NOT FUNCTIONING.	These may be worn, damaged, stuck or loose. Examine, clean, tighten and lubricate or replace.

(c) Testing

If a low pressure cut out is suspected of being faulty the following tests can be carried out.

Plant not running. Remove the control front cover and 'short out' the electrical connections. If the compressor starts, the control is faulty.
Plant not shutting down. Reduce the evaporating pressure, by restricting the liquid flow into the evaporator, to a point well below CUT-OUT. If the compressore continues to run the control is faulty.

The first test 'shorting out' can be carried out in an identical way on the thermostat and high pressure cut out. It is a quick, easy way of determining which control has shut down the plant.
The second test can also be used for the other controls except for THERMOSTATS. (The sensing element should be cooled by spraying with liquid refrigerant.)

High pressure cut outs. The condensing pressure should be CAREFULLY raised. Switch off the condenser fan or in some other way restrict the flow of air or water onto the condenser.

6 SOLENOID VALVES

These are usually fitted in the liquid line, thermostatically controlled, to shut off refrigerant flow or in the condenser cooling water pipe, to shut off the water during the compressor off cycle. They are also used for

Closing off evaporator outlets (multi-circuit operation),
Reversing refrigerant flow (de-frosting and heat pump applications),
Controlling the flow of brine and other liquids.

They must be correctly installed. Unless designed for horizontal installation, solenoid valves must always be mounted in a vertical position with the coil on top. They must be installed in line with the direction of flow, usually shown by an arrow on the valve body.

Faults	*Remedy*
Valve fails to open	Ensure valve is mounted level on the pipe. Check that power is available at terminals. Check that connections are clean and tight. Dismantle and check that the plunger is not sticking. Test the coil for continuity.
Valve does not close fully	Dismantle and clean valve, valve seat and plunger. If valve seat is worn re-new or fit a new valve.
Valve is noisy in operation	Valves that 'chatter' are almost always not installed properly. Check that the valve is level and that the electrical connections are sound.

7 PRESSURE OPERATED WATER VALVES

These valves regulate the flow of condenser cooling water to maintain a constant condensing pressure.

As the plant load increases, the condensing pressure also increases and this causes the valve to open and increase the cooling water flow rate. During winter, or when the load is low, the condensing pressure decreases and the cooling water flow rate is reduced. When the compressor is shut down, the condensing pressure decreases rapidly and, with a well adjusted valve, the flow of water is cut off completely. This results in a water saving.

Faults	*Causes*
Insufficient water flow	Valve requires adjusting to increase flow. Inlet strainer requires cleaning. Valve seat dirty and requires cleaning. Bellows leaking and must be replaced.
Water flow too high or does not close off on compressor shut-down	Valve requires adjusting to reduce flow. Water pressure too high and needs reducing. Large particles of dirt between the valve and its seat need flushing out. Valve seating worn and requires replacement. Return spring broken.

SECTION G

SERVICING TECHNIQUES

1 BURN OUT REPAIR

Action following the burn out of a hermetic or semi-hemetic compressor motor.

1. Disconnect the electrical supply.
2. Clean the outside of all surfaces where it is intended to break into the system.
3. Discharge the refrigerant to outside air. Test the accompanying oil for acid.
4. Remove the compressor assembly. If the oil is acidic,. use goggles and rubber gloves.
5. Inspect compressor inlet and discharge for carbon deposits. Test for acid. If system is contaminated, proceed to 6. If clean proceed to 7.
6. Flush the system with R11 to remove contaminants or fit a high capacity burn out drier in the compressor SUCTION pipe.
7. Install the replacement compressor and test for leaks.
8. Replace the liquid line drier and then evacuate the system.
9. Charge and test for leaks.
10. If a suction line drier has been fitted, remove after 24 hours and check the circulating oil for acid and fit new drier until all acid traces have been removed.

2 CAPILLARY TUBE CLEANING

Blocked capillary tubes are frequently cleared of obstructions rather than replaced. The connection to the evaporator is too inaccessible for a joint to be made.

1. Discharge the refrigerant to the outside air.
2. Clean the outside surfaces and then remove, by heating the soldered/brazed joints, the filter-drier and the capillary tube inlet.
3. Attach the capillary tube cleaner (see SECTION B, Fig. 5) to the open end.
4. Build up sufficient pressure to clear the obstruction.
5. After the tube has been cleared flush it with liquid refrigerant.
6. Install a new filter-drier and re-make all connections.
7. Evacuate the system, charge and test for leaks.
8. If it is suspected that wax deposits caused the obstruction, change the oil in the system.

3 CHARGING THE SYSTEM

Systems fitted with service valves.

(a) Vapour (gas) charging (see Fig. 16) commercial systems

1. Fully back seat the compressor suction service valve and remove the flare nut and bonnet.
2. Connect a charging line to the outlet (vapour) connection of the refrigerant cylinder.
3. Purge all air and contaminants from the charging line by cracking the cylinder valve slightly. Direct the escaping refrigerant downwards onto the floor. Make sure that vapour NOT liquid is being withdrawn from the cylinder.
4. Connect the open end of the charging line to the service valve. Leave this connection 'finger tight'.

5 Remove any air remaining in the charging line from the loose connection at 4 by cracking the cylinder valve again. When refrigerant is heard escaping at the connection, tighten it.
6 Open the cylinder valve fully and the suction service valve to the mid-way position.

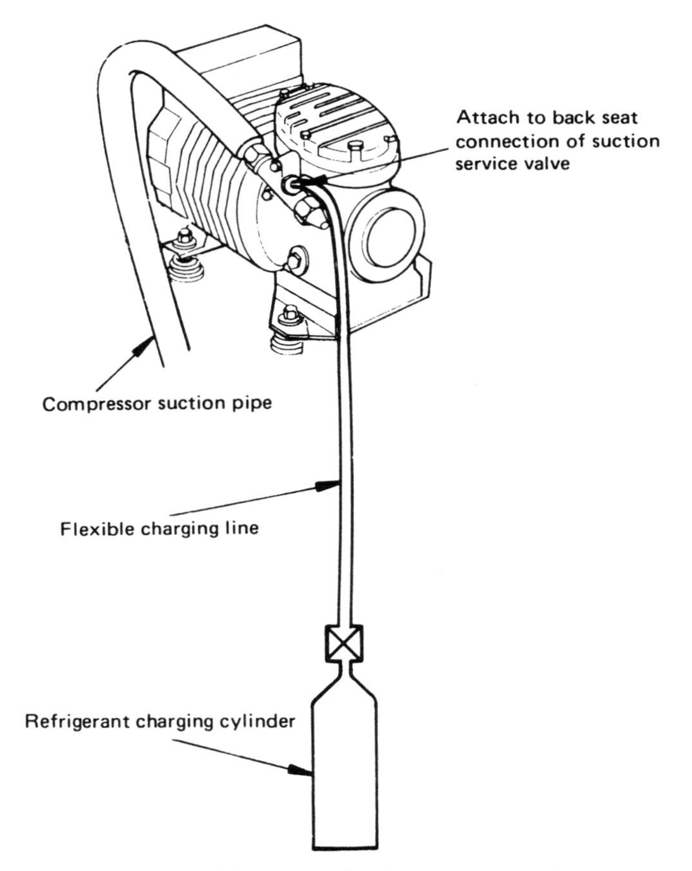

Figure 16. Vapour charging arrangement

7 Run the compressor, and condenser fan, until the liquid sight glass shows a clear stream of liquid i.e. no vapour bubbles are present.
8 Fully close the cylinder valve and after a few seconds fully back seat the suction service valve.
9 Remove the charging line and re-plug the ends. Replace the flare nut and bonnet.
10 Set the suction service valve to the normal running position.

(b) Liquid charging (see Fig. 17) industrial and some larger commercial systems

1 Connect a charging line to the outlet (liquid) connection of the refrigerant cylinder.
2 Purge this line as described in paragraph (a) 3. Make sure that liquid is being withdrawn from the cylinder. Wear goggles and avoid liquid contact with the skin.
3 Connect the other end of the charging line to the liquid charging valve. Remove any remaining air as described in paragraph (a) 4 and 5.
4 Fully open the cylinder outlet valve.
5 Slowly CLOSE the liquid receiver outlet valve and OPEN the liquid charging valve until liquid starts to flow out of the cylinder and into the system.
6 When sufficient refrigerant has been added to the system, fully close the cylinder outlet valve. After a few minutes fully close the liquid charging valve and fully open the liquid receiver outlet valve.
7 Remove the charging line CAREFULLY. It may still contain liquid. Use goggles and rubber gloves.
8 Test for leaks at all valves which have been operated. Cap off as previously found. Ensure plant is operating satisfactorily.

For systems that are NOT fitted with service valves follow the procedure at (a). A connection will have to be made on the suction side of the compressor, as described below under SEALED SYSTEM CONNECTION.

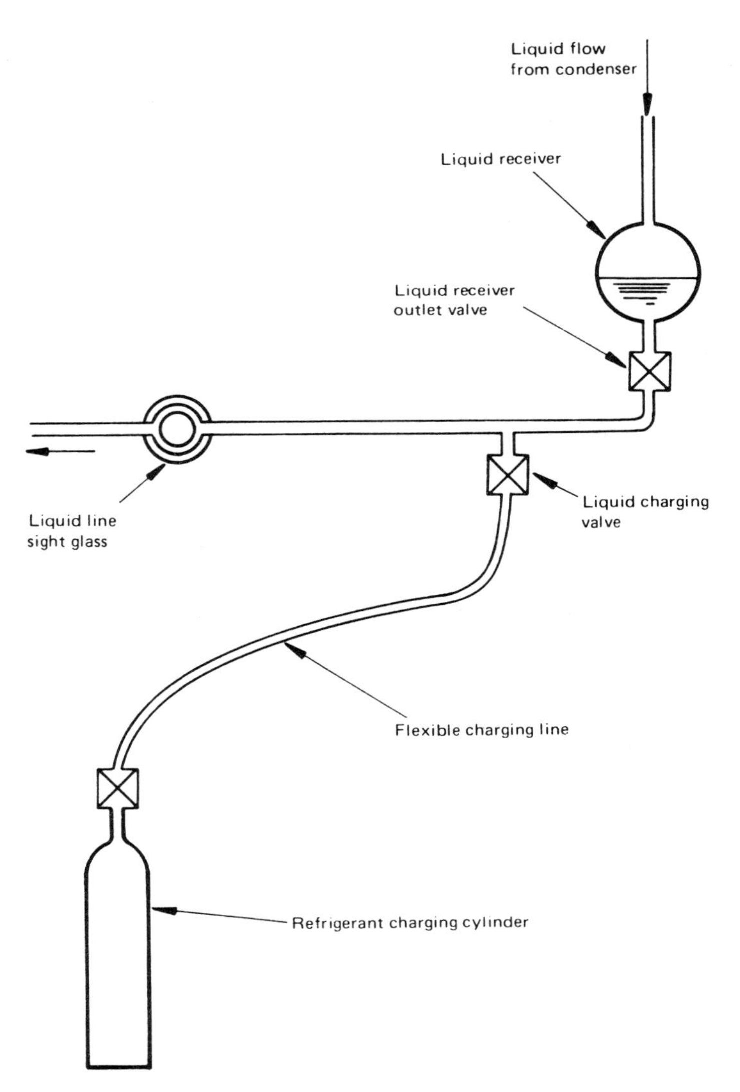

Figure 17. Liquid charging arrangement

(c) Checking that the system is correctly charged see SECTION H INSTALLATION TECHNIQUES

4 COMPRESSOR WORK

(a) Changing or replenishing the oil (commercial plant)

Examine the compressor oil at intervals. If it is discoloured, contamination is taking place. The oil should be clear. Check the level of the oil. Contaminated oil and low oil level are common causes of compressor failure.

For compressors fitted with filling/checking plugs, adopt the following procedure.

1 Fully front seat the compressor suction service valve.
2 Run the compressor until the suction pressure gauge indicates zero. The crankcase is at atmospheric pressure. Re-set the low pressure cut out as necessary.
3 Switch off the compressor. The suction pressure will usually rise as dissolved refrigerant is boiling out of the crankcase oil. Re-start the compressor to remove this vapour and repeat the procedure until all crankcase pressure has been relieved.
4 Switch off the compressor and front seat the delivery service valve.
5 Remove the oil filling plug and inspect the oil. If it is dirty adopt the following procedure. If it is clean but the level is low, proceed to 9.
6 Insert a suitable length of plastic tube through the plug hole. Where possible use a tube that is a tight fit in the hole, otherwise wind PTFE, or similar, tape around the outside of the tube to effect a joint.
7 Ensure that the end of the tube inside the compressor reaches the lowest point in the crankcase. Place the other end of the tube into a container of sufficient capacity.
8 Crack the suction service valve open and allow a small quantity of vapour to force the oil out of the crankcase.
9 Add oil of the correct type and quantity through the plug hole and then replace the plug securely.

10 Remove air and moisture by 'pumping down' the compressor. See paragraph (b) below.
11 Re-set the service valves to the normal run position.
12 For industrial plant, fitted with oil sight glasses and charging facilities, follow the manufacturer's instructions.

(b) Pumping down the compressor

Whenever a compressor is opened to the atmosphere, for repair or maintenance work, air and associated moisture will enter. This MUST be removed. Where possible, it should be carried out with a vacuum pump (see Section H EVACUATION OF THE SYSTEM).

The compressor can however frequently be used to draw a vacuum upon itself. The procedure termed 'PUMPING DOWN' is as follows.

1 Fully front seat both the suction and delivery service stop valves.
2 Remove a connection (such as a delivery gauge or high pressure cut out) fitted between the delivery from the compressor and the closed delivery service stop valve.
3 Re-set the low pressure cut-out to its lowest (minimum) setting.
4 Start the compressor and make sure that the air inside is being discharged through the open connection provided at paragraph 2.
5 Run the compressor until a vacuum of at least 20 in (0.50 m) of mercury is obtained.
6 Stop the compressor. Crack the suction valve slightly to fill the compressor body with refrigerant vapour from the suction pipe.
7 Re-run the compressor as in paragraph 5.
8 With the compressor still running, close off the connection removed at paragraph 2 and examine the delivery pressure gauge. It MUST indicate a zero pressure. Any pressure, above atmospheric, shown on this gauge indicates that air and/or

moisture has not been removed from the compressor and the pumping down must be repeated.

9 When pumping down has been satisfactorily completed, set both valves and the low pressure cut-out to their normal running positions.

10 PRECAUTIONS: Not all compressors are suitable for pumping to low vacuums. Consult the manufacturer's instructions and adhere rigidly to them.

At low vacuum there is a tendency for oil to be drawn up from the crankcase. This will damage the compressor and must be avoided. It is accompanied by a pounding noise, listen and switch off the compressor if it occurs.

(c) Pumping efficiency tests

To test whether the suction valves are seating well.

1 Fully front seat the suction service stop valve.
2 Re-set the low pressure cut out to its lowest (minimum) setting.
3 Start the compressor and find what vacuum it will produce in five minutes. The extent of this vacuum indicates the condition of the suction valves (see Fig. 18).
4 Switch off the compressor and observe the rate at which the vacuum is lost. If the pressure, as shown on the suction gauge, rises rapidly this is a further indication that the suction valves are not seating properly.
5 PRECAUTIONS: Avoid possible damage by following the instructions given above under 'PUMPING DOWN THE COMPRESSOR' paragraph 10.

To test whether the delivery valves are seating well

1 Run the plant until a normal condensing pressure is obtained.
2 Switch off the compressor and fully front seat the delivery service stop valve.
3 If the delivery valves are seating well the delivery pressure, as shown on the gauge, will remain steady. Conversely if the

A. Indicates a vacuum of less than 10" (254mm). This is *not* acceptable.

B. Indicates a vacuum of at least 20" (508mm). For most compressors, this is acceptable.

C. Indicates a vacuum of approximately 25" (635mm). This is very satisfactory. The suction valves are seating well.

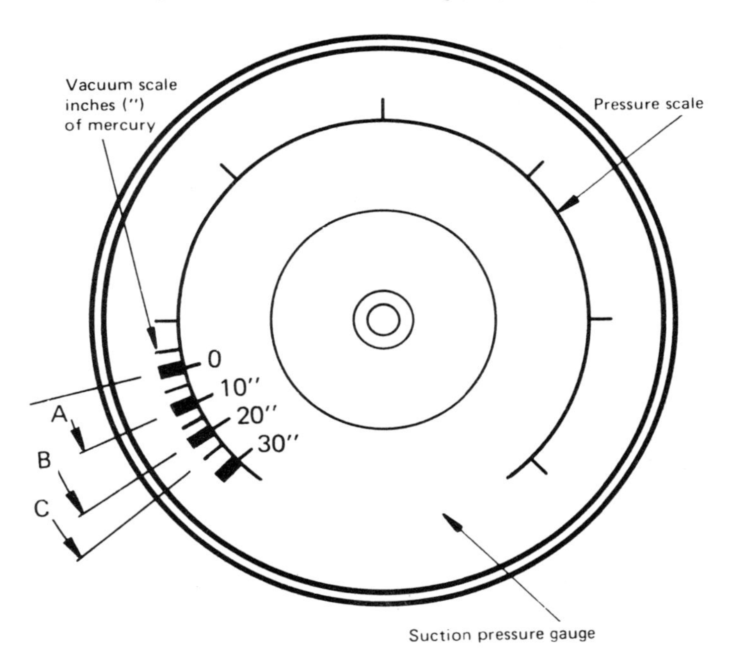

Figure 18. Testing the condition of the compressor suction valves

delivery valve reeds are leaking badly the pressure will fall rapidly. The rate of loss of pressure will indicate how well the delivery valves are seating in the same way as the rate of rise of pressure showed the condition of the suction valves in the previous test.

4 On completion make sure that the compressor delivery service stop valve is opened to its normal running position.

(d) Replacing a faulty valve plate (commercial compressors)

1 Front set the suction service valve and remove all refrigerant from the compressor body. See paragraph (a) 1 to 4.
2 Thoroughly clean with kerosene, or similar, the outside of the cylinder head, valve plate and upper part of the compressor body.
3 Remove the cylinder head bolts and place them in a clean container.
4 Remove the cylinder head and valve plate carefully. Try not to damage the gaskets as they can be re-used. Do not use excessive force or any metallic tool to remove these components. A cylinder head that 'sticks' can be loosened by tapping the lug provided with a piece of hard wood.
5 Place clean rag into the cylinder bores to prevent dirt from entering. If the gaskets have stuck, remove all particles from the cylinder head and compressor body with a scraper. Be careful not to damage either of the surfaces and make sure that they are thoroughly clean.
6 Remove the rag and clean the cylinder bores.
7 Apply a light film of oil to the surfaces of the gaskets then re-assemble in the following order:

   (a) Position gasket onto compressor body.
   (b) Examine replacement valve plate. Ensure it is the right type, complete and undamaged and then position it onto the gasket the correct way up.
   (c) Position a second gasket on the top surface of the valve plate.
   (d) Re-fit the cylinder head. Ensure that the holes of all the above components are correctly lined up then replace the cylinder head bolts which should be lightly oiled and screwed into position by hand.

8 Tighten the bolts carefully and evenly. Follow any instructions given by the manufacturer regarding sequence of tightening and torque applied. DO NOT overtighten. If leakage occurs, the bolts can be re-tightened. Bolts that are over-

tightened and break are difficult to remove.

9 On completion, remove all air and moisture that has entered the compressor by pumping down, as described in paragraph (b).

5 EXPANSION VALVE (THERMOSTATIC) SERVICING

This type of valve is designed to ensure that the compressor suction vapour is slightly superheated since superheated vapour can-

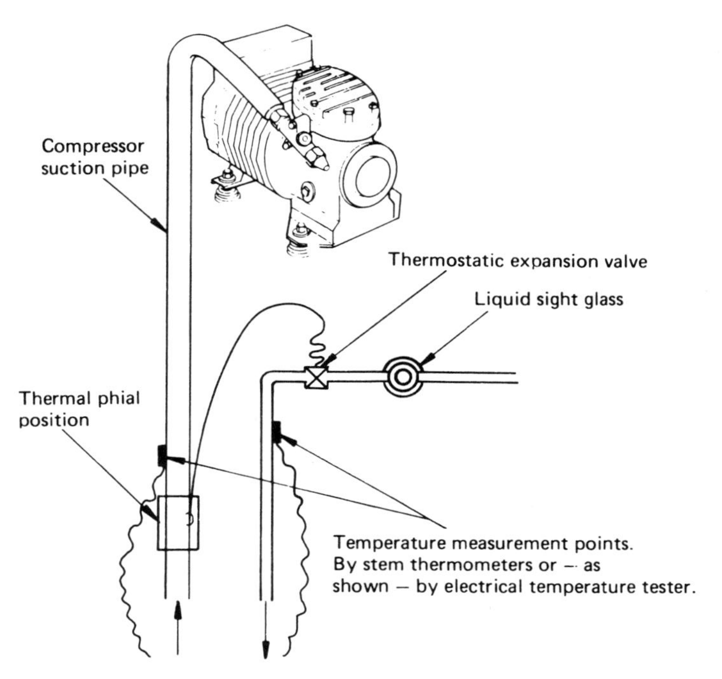

Figure 19. Measurement of suction superheat

not carry liquid refrigerant back to the compressor and cause hydraulic damage.

(a) Measurement of suction superheat

1 Attach stem type thermometers, or the probes of a temperature analyser, to the expansion valve outlet and thermal phial positions as shown in Fig. 19.
2 Ensure that these thermometers are securely fitted, in good thermal contact with the pipe and that they are well insulated from the effects of the surrounding air. Try to find the true temperature of the pipe at the point of measurement.
3 Start the plant and when stable running conditions have been obtained take the two thermometer readings. The suction superheat is the difference between the temperature at the phial and the expansion valve outlet. For most plants, the phial temperature should be 10 to 15 °F (5 to 8 °C) above the evaporator outlet temperature.

(b) Suction superheat too low

The valve is passing too much refrigerant and the suction pipe, beyond the thermal phial position, is frequently sweating or frosting. Check in the following order:

1 That the thermal phial is not loose on the pipe, and re-tighten if necessary.
2 That the phial is not in a flow of warm air, and insulate the outside of the phial if necessary.
3 The adjustment of the valve. Re adjust to increase the superheat if necessary.
4 That the valve orifice is the correct size. Remove, examine and compare with supplier's recommendation. A smaller orifice may be required.

5 That the needle valve is not stuck open. Replace the valve if necessary.

(c) Suction superheat too high

The valve is not passing sufficient refrigerant. The evaporator coil will not be frosting over its full length and the compressor will be running for longer periods and hotter than usual.
Check in the following order:

1 The filter at the liquid inlet to the valve which may be blocked with dirt and restricting the flow.
2 Moisture frozen at and blocking the valve seat. First, examine the liquid line drier, where fitted. Then warm the valve to melt the ice. Fit or change the drier to absorb the moisture released into the system.
3 Wax deposits on the valve seat. The compressor oil is unsuitable for the evaporating temperature. Change the valve and also the oil, for one suited to low temperature work.
4 The adjustment of the valve. Re-adjust to reduce the superheat.
5 The valve orifice size. It may be too small. See paragraph (b) 4.
6 That the needle valve is not stuck shut. Replace the valve if necessary.
7 That the thermal phial has not lost its charge. If necessary, remove the phial from the pipe and warm it by hand. The valve should immediately open and the suction pressure rise.

(d) Electrical circuits (servicing) see SECTION H INSTALLATION TECHNIQUES

(e) Evacuation of the system see SECTION H INSTALLATION TECHNIQUES

6 HERMETIC COMPRESSOR MOTORS (STARTING PROBLEMS)

Sometimes these motors will not start, even though electrical tests show no fault. This usually occurs after the motor has not run for some time and is due to copper plating effects, small dirt particles or an excessive accumulation of liquid refrigerant in the shell. The following methods can be used to free the motor.

1 By-pass the starting relay and connect the motor terminals direct to the power supply.
2 Reverse the direction of rotation of the motor. With a capacitor start motor, install the capacitor in series with the running winding.
3 Use a higher than normal voltage. This must only be carried out for a very short period to avoid damage to the motor windings.

Leak detection methods see SECTION H INSTALLATION TECHNIQUES

Non condensable gas in the system see SECTION H INSTALLATION TECHNIQUES

7 REPAIRING LEAKS

1 Remove all refrigerant from the section to be repaired. If necessary completely discharge the system. DO NOT solder or braze a system containing refrigerant as the heat may break down the refrigerant.
2 Use the lowest possible temperature to obtain a satisfactory repair. High temperatures form scale on the inside surfaces, which contaminate the system.
3 Find what metal is used at the point of leakage. Scrape the surface: steel is grey and magnetic; copper is dull red and non-magnetic; aluminium is white and non-magnetic. Test

with a small magnet. Use the most suitable jointing technique. (See SECTIONS C and D).

4 A leakage from brazed or silver soldered connections can usually be cured effectively by cleaning, applying flux and reheating.
5 If leakage from a flared joint cannot be prevented by tightening, the joint should be disconnected and a new flare made on the pipe. Fit a new flare nut and fitting and soften the end of the pipe by annealing.
6 If the repair necessitates 'breaking into' the system ensure that all contamination is removed by purging. See below.

Purging the system see SECTION H INSTALLATION TECHNIQUES

8 SEALED SYSTEM CONNECTION

Hemetic systems do not usually have connections for carrying out servicing techniques such as charging and pressure measurement. Access is usually made by fixing piercing (LINE TAP) valves to the suction and discharge sides of the compressor. The procedure is as follows:

1 Choose a length of pipe that is straight, round and has no scratches, marks or dents on it. Make sure that there is sufficient space to operate the valve attachment.
2 Clean the pipe carefully and then select a valve of the correct size. Apply a small quantity of refrigerant oil to the outside of the pipe.
3 Fit the valve on the pipe as shown in Fig. 20. Make sure that the sealing washer is in place and that the piercing valve is screwed FULLY OUT.
4 Tighten the clamping screws evenly until the valve is fixed rigidly to the pipe.
5 Remove the sealing cap and gasket and replace it with the service valve attachment. Tighten this down firmly to prevent leakage.

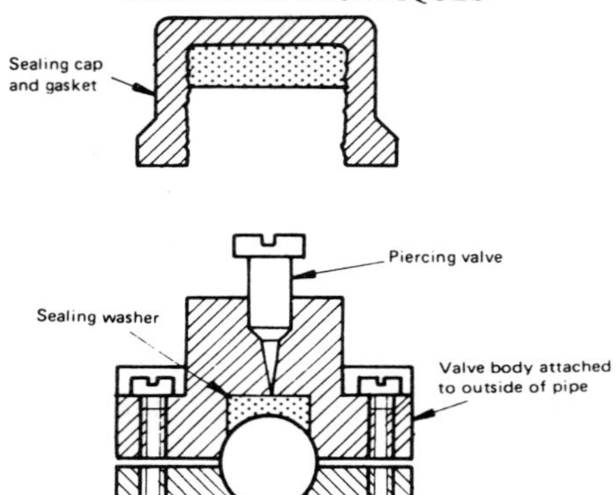

Arrangement before the pipe has been pierced.

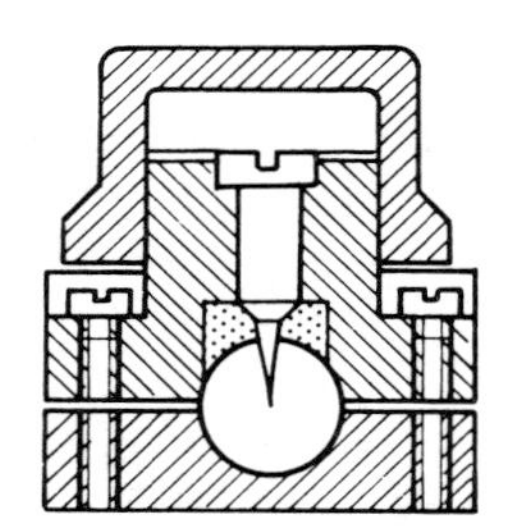

Arrangement after the pipe has been pierced

Figure 20. Fitting a piercing or 'line tap' valve

6 Depress the handle of the service valve until the screwdriver type attachment engages with the slot in the top of the piercing valve.
7 Turn the handle of the service valve clockwise until the piercing valve has made a hole in the pipe.
8 Turn the handle anti-clockwise and vapour will pass from the hole in the pipe, through the valve and out of the connection to the gauge fitted.

9 On completion of the servicing work replace the piercing valve into the hole in the pipe. Repeat the procedure at paragraphs 6 and 7.

10 Remove the service valve and replace the sealing cap and gasket. Test carefully for leaks.

Starting the system

Starting analyser

Transferring and handling liquid refrigerant

See SECTION H INSTALLATION TECHNIQUES

SECTION H

INSTALLATION TECHNIQUES

Installation of a new plant is usually carried out in the following sequence.

Installation of main components
Pipe erection work
Installation of valves and controls
Electrical installation work
Test system for leakage
Evacuate the system
Charge the system
Start and test the plant

The following notes, in alphabetical order, together with those in SECTION G will be found helpful when installing a new plant or during the major overhaul of an existing one.

Charging the system see SECTION G SERVICING TECHNIQUES

1 CHECKING THE CHARGE

The charge in the system must be correct. If a plant is undercharged, it will be unable to cool satisfactorily. If it is overcharged, the compressor discharge pressure will be abnormally high. The following types of plant are particularly sensitive to over/undercharge. (The charger in the system must be correct.) Plants fitted with either high pressure float valves or capillary tubes and those NOT fitted with liquid receivers.

For effects of over and under charge see also SECTION K MECHANICAL FAULT FINDING. Methods of ensuring that the correct charge of refrigerant is admitted to the plant are listed below.

1 By means of the charging cylinder (See SECTION G CHARGING THE SYSTEM). Charge the precise quantity of refrigerant into the system recommended by the supplier.
2 On large plants fitted with test cocks, by checking the liquid level in the receiver. The lower cock when opened should release liquid, the upper vapour.
3 On plants fitted with liquid line sight glasses, by the state of the refrigerant. See SECTION G, CHARGING THE SYSTEM.
4 By shutting down the plant and releasing vapour from the liquid receiver, The boiling action which occurs lowers the liquid temperature as latent heat is extracted. The level of liquid in the receiver can then be seen by the 'sweat' line that appears on the outside.

Electrical installation work see SECTION E

2 ELECTRICAL CIRCUITS (SERVICING)

1 Ensure that equipment is correctly connected, as shown in the wiring diagram.
2 Ensure that the cables are adequately sized for the current carrying capacity.
3 Ensure that there is power at the equipment and that the voltage is correct.
4 Ensure that the current drawn by the equipment (when the electric motors are on load and up to speed) conforms with the supplier's rating.
5 If any motor will not run. TURN OFF THE POWER. Check the circuits for continuity with an ohmmeter. Find the break in the circuit continuity by measuring sections between terminals or connections.
6 Do not attempt to start an electric motor unless there is an effective overload cut-out in the supply circuit.

3 EVACUATION OF THE SYSTEM

The method used to remove air and moisture from a system depends upon the equipment and time available and the type of system and the degree of contamination that is initially present and which can, after evacuation, be tolerated in the system.

The following methods are progressively more effective in removing contamination:

(a) Triple evacuation

Used when there is no free moisture present in the system, or on AMMONIA plants which can tolerate moisture.

The method is fast, needs no special equipment, and is suitable for most industrial work.

1 Use the system compressor to draw the best possible vacuum without risk of oil being drawn up from the crankcase. See SECTION G PUMPING DOWN THE COMPRESSOR. DO NOT use hermetic compressors for this method. They rely on the flow of cold suction vapour to prevent overheating of the motor.
2 Admit refrigerant to the system, sufficient to bring the pressure back to atmospheric.
3 Pump down the system again as in paragraph 1.
4 Admit refrigerant as in paragraph 2.
5 Repeat this procedure THREE times.

(b) Multiple evacuation

1 With a high capacity vacuum pump connected to both the high and low pressure sides of the system (as shown in Fig. 21), draw a vacuum of 2 Torr. (2 mm of mercury above a perfect vacuum).
2 Admit refrigerant as in paragraph (a) 2 above.

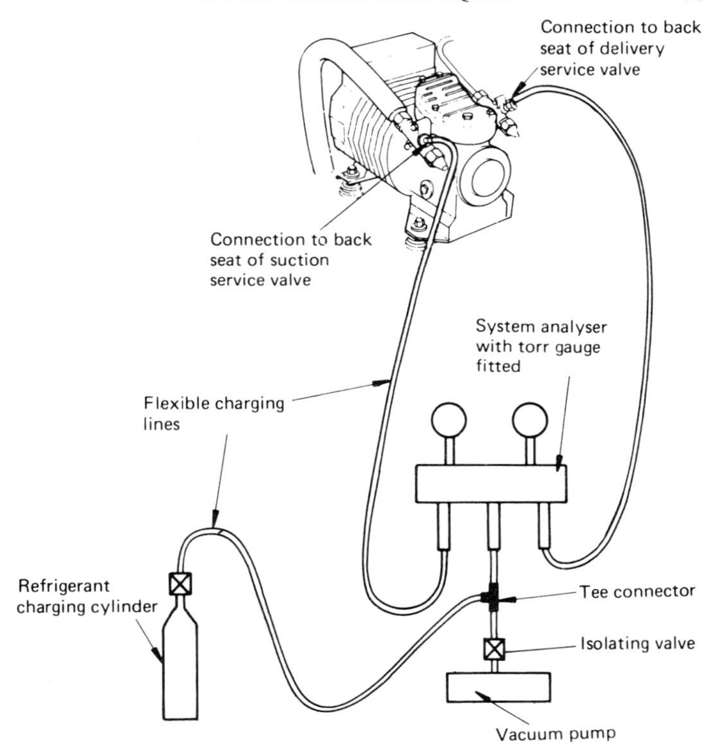

Figure 21. Arrangement for evacuating the system

3 Evacuate the system again to a vacuum of 2 Torr.
4 Close the valve connecting the vacuum pump to the system and then switch off the pump.
5 Leave the system standing under vacuum for as long as circumstances permit. Any rise in pressure, provided that the system is leak tight, is due to water vapourising at the low pressure. Run the vacuum pump to remove it until the required vacuum of 2 Torr can be maintained.

(c) Deep vacuum de-hydration

This method relies upon the principle that if a low enough vacuum is produced ALL moisture will evaporate and be removed from the system. This is the recommended method for most commercial work.

A pump capable of producing very low vacuums and an accurate method of measurement is required.

1 Connect the vacuum pump as shown in Fig. 21, except that on large systems the use of a cold trap, as shown in Fig. 22, is recommended to give a quicker rate of moisture removal.

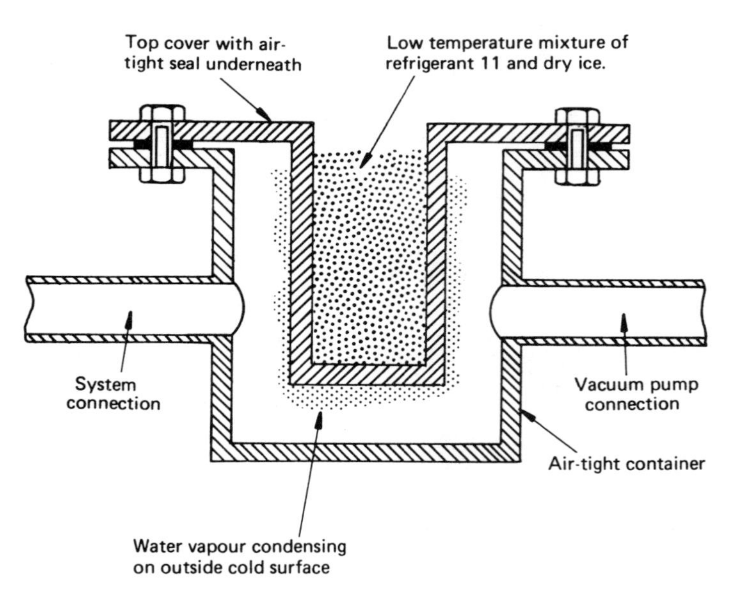

Figure 22. 'Cold trap' for moisture removal

2 Run the pump until a vacuum of 0.1 Torr is obtained.

3 Close the valve, switch off the pump and leave the system

standing under vacuum. Repeat as necessary until the required vacuum can be maintained.

Whichever method is used ensure that connections are provided to enable the vacuum produced to be broken, and the system charged, with clean, dry refrigerant.

4 INSTALLATION AND LOCATION OF MAIN COMPONENTS

Compressor sets

Install on cork mats or anti-vibration mountings, particularly in air conditioning work, to avoid transmission of noise.

Ensure that the floor loading limit is not exceeded.

Position where there is good access for maintenance, repair and possible replacement.

Check that the position selected will be cool and dry in ALL seasons and will provide good ventilation.

Where possible locate the equipment where it will remain clean, not liable to damage or be seen by the public.

If the compressor is installed on an anti-vibration mounting flexible connections MUST be provided on all refrigerant, oil and water connections.

When installing

(a) Dismantle as little as possible, keep all components clean and re-assemble as soon as possible.
(b) Ensure that all components are level, securely held by adequately sized foundation bolts and that externally driven compressors are correctly lined to the driving motor.
(c) Driving belts must be correctly tensioned and replaced, when necessary, by a complete matching set.
(d) Do not release the compressor 'holding charge' of refrigerant until all of the piping work has been completed.
(e) Follow carefully the instructions for electrical connection and testing given in SECTION E.

Condensers

AIR COOLED. Position close to the evaporator and, where possible, slightly above rather than below.

Position in a cool, clean, dry, well ventilated location, protect from excessive sun heat, avoid air re-circulation possibility, take advantage of the prevailing wind during summer, maximum load, conditions.

Allow for access, noise problems and possibility of damage, as with compressor sets.

WATER COOLED. Position generally as for air cooled but effects of wind and air re-circulation are not applicable and those of sun heat and dirt less important.

Ensure that there is sufficient space for end cover removal and tube replacement.

Take suitable precautions at the installation stage to avoid damage due to freezing in the winter.

Evaporators

FORCED DRAUGHT

Position where there is good access to the unit and its auxiliaries (expansion valve and defrost equipment) and, in particular, the fan motor.

Ensure that the flow, and return, of air is not obstructed.

Arrange to support the unit such that noise and vibration from the fan will not be transmitted, excessive heat is not conducted from outside through the supports, and corrosion (due to condensation) of the supports is minimised.

LIQUID CHILLING

Locate and install generally as described under water cooled condensers.

Controls See SECTION F.

5 LEAK DETECTION METHODS

Refrigerants are selected to suit the evaporating temperature, so that under normal operating conditions the pressure at any point in the system is ABOVE atmospheric and any leakage will be of refrigerant outwards.

In the absence of leak-detecting equipment, a water-soap solution brushed over the suspected area is satisfactory as bubbles appear.

Fluorocarbon refrigerants

These have a high rate of leakage, particularly from packed glands.

Leaks are indicated by the presence of oil on the outside of a joint, as oil circulates around the system and escapes with the refrigerant.

The fluorocarbons are heavier than air, so test for leaks on the underside of joints.

Expanded insulating materials frequently use this refrigerant as the expanding agent and detection devices show a leak continuously. Use a water-soap solution.

Detection Methods

(a) *The halide lamp.* Similar in construction to a blow lamp using a fuel such as alcohol or a brazing torch, using propane. The lamp burns with a clear flame until leaking refrigerant, drawn in through a search tube, passes across a copper element, when the flame turns a light green colour.
(b) *Electronic type.* These operate because of the difference in electrical resistance between air and refrigerant. A search tube is used to draw samples of air from suspect areas and pass them across a measuring device. The presence of refrigerant is indicated by a visual or audible alarm, or both.

Most instruments are battery operated and extremely sensitive, and will indicate very small leaks.

Handle them carefully, taking care not to drop them. Avoid using in high concentrations of refrigerant.

Move the detecting tip slowly, avoid draughts and keep the tip clean.

Ammonia

Leaks are detected by burning SULPHUR which combines with ammonia to give a dense white smoke.

Sulphur candles, wood slivers coated with sulphur, are obtainable which when lit and passed around a leaking joint show the high pressure ammonia escaping as a jet of white smoke.

Although ammonia is a flammable refrigerant, there is no danger associated with this method of leak detection provided that the concentration is low enough to work in. DO NOT use any form of flame if the concentration of ammonia is so great that breathing is uncomfortable.

6 NON-CONDENSABLE GAS IN THE SYSTEM

Any gas that will not condense remains in the high pressure side of the system. Its effect is to raise the delivery pressure, which in turn increases the rate of wear of the compressor and its power consumption (See also SECTION K).

The presence of non-condensable gas (NCG) is usually due to one of the following.

1 Air left in the system after installation or repair, i.e. evacuation was not carried out properly.
2 Air drawn in through leaking joints on the low pressure side of the system.
3 Chemical breakdown of refrigerant or oil, due to high delivery temperatures.

(a) Testing for non-condensable gas

If the delivery (condensing) pressure is above normal and NCG is suspected as being the cause:

1 Switch off the compressor and, where fitted, the evaporator circulating fan or pump.
2 Run the condenser fan, or water circulating pump, until the delivery pressure stabilises, i.e. the pressure will not fall any lower.
3 The pressure shown on the delivery gauge is termed 'the standing head pressure'. It should correspond to the vapour pressure of the refrigerant at the surrounding air (ambient) temperature.
4 If NCG is present in the system, the standing head pressure will be greater than the pressure described in paragraph 3 by an amount equal to the excess pressure created by the NCG. See Fig. 23.

Both gauges show the 'standing head pressure'.
Compressor is *not* running.

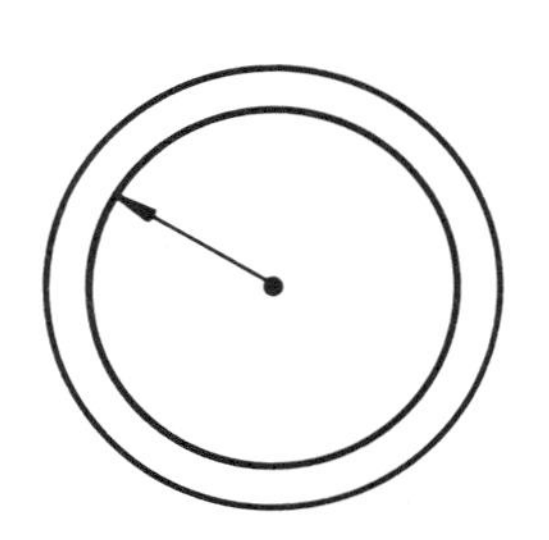

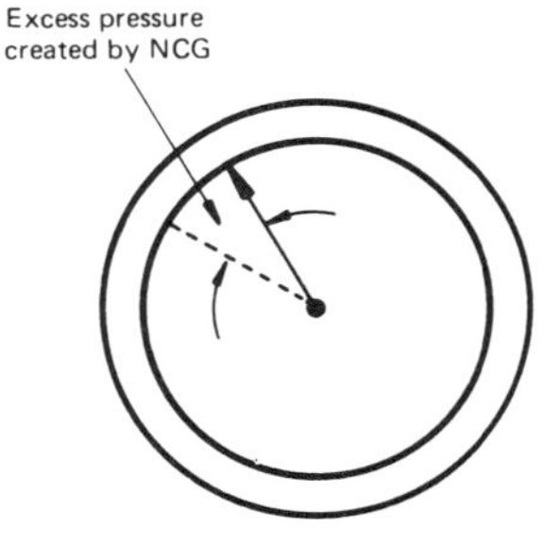

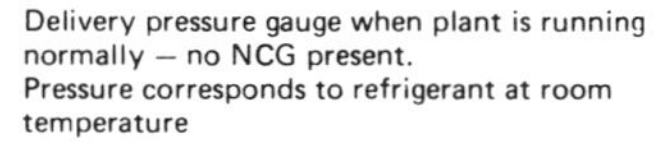

Delivery pressure gauge when plant is running normally – no NCG present.
Pressure corresponds to refrigerant at room temperature

Delivery pressure gauge when NCG is present in the system.
Pressure is higher than refrigerant at room temperature.

Figure 23. The effect of non-condensable gas

(b) Removal of non-condensable gas

1 Switch off the condenser fan and calculate the pressure being exerted by the NCG.
2 Purge from the system one half of the excess pressure. Purge from the highest point on the high pressure side of the system SLOWLY.
3 Wait five minutes for the delivery gauge to stabilise and then calculate the remaining excess pressure.
4 Purge one half of this remaining excess pressure, as in paragraph 2.
5 Continue this procedure. Purge one half of the excess pressure, followed by a five minute wait for the pressure to stabilise, until the excess pressure is approximately 1 psi (0.07 bar) only. This small remaining pressure should then be purged from the system until the delivery gauge pressure corresponds to the ambient temperature.
6 *Note.* Industrial plants are frequently fitted with automatic purgers to remove NCG from the system.

7 PIPE INSTALLATION WORK

1 Ensure that the correct size of pipe is used. The size of connection provided at the component is not always suitable for the connecting pipe. Small pipes result in a pressure drop, due to friction, which reduces the compressor capacity, increases the power consumption and restricts the liquid flow.

Large suction pipes have too low a velocity for the suction vapour to carry the circulating oil back to the compressor (fluorocarbon refrigerants).
2 Position all pipes so that they are accessible for repair or replacement and unlikely to get damaged.
3 Pipe-work should have a neat appearance. All bends should be smooth and straight, with runs either horizontal or vertical.

4 Adequate supports should be provided and pipes passing through walls, floors etc. should incorporate a sleeve. See Fig. 24.

5 Refrigerant quality pipe for commercial work is supplied clean and dry with the ends sealed. Do not open a length of pipe until it is required and re-seal the length afterwards. Try to plan the work so that a long length of pipe is not opened towards the end of a working day.

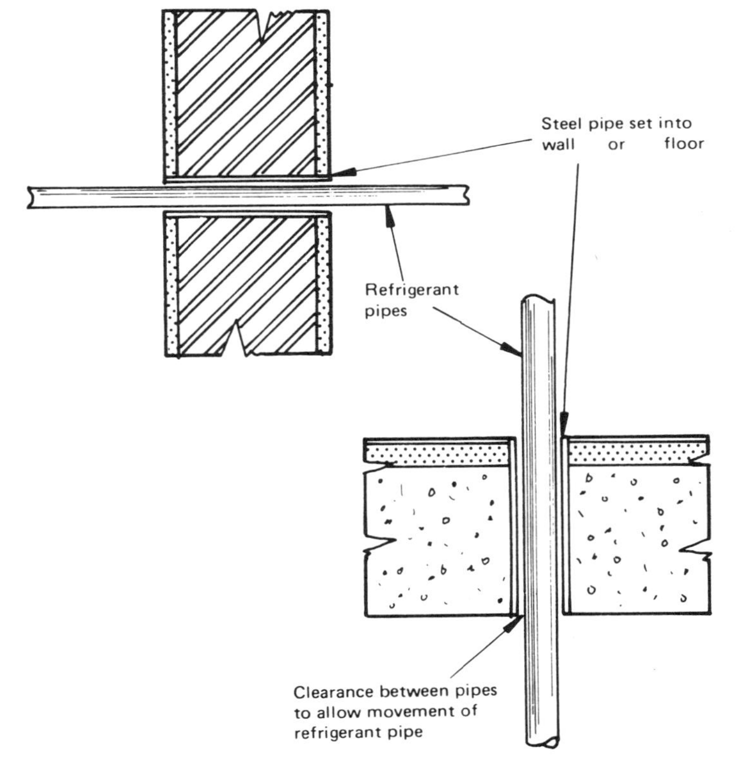

Figure 24. Refrigerant pipe sleeves

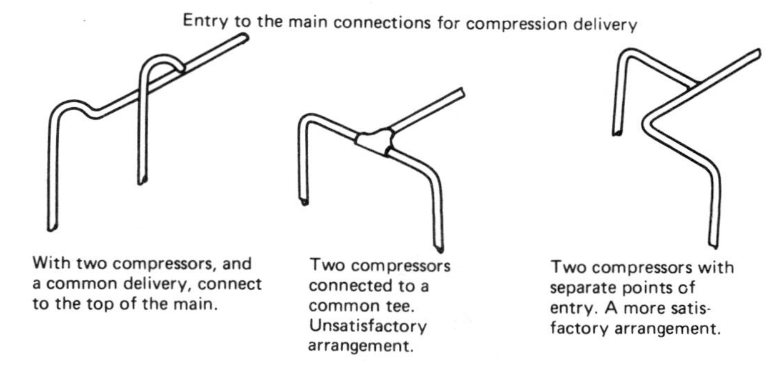

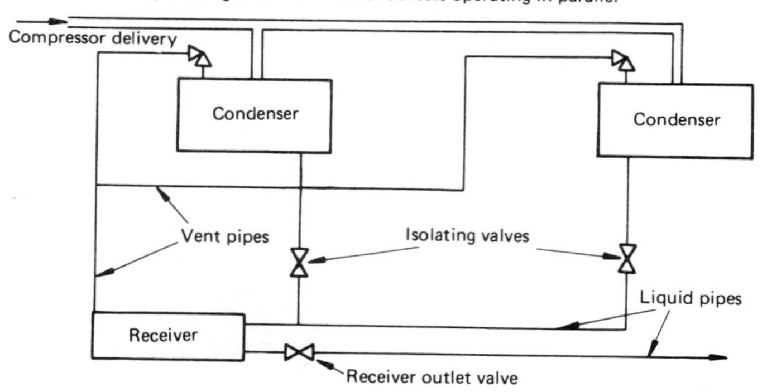

Figure 25. Refrigerant pipe arrangements

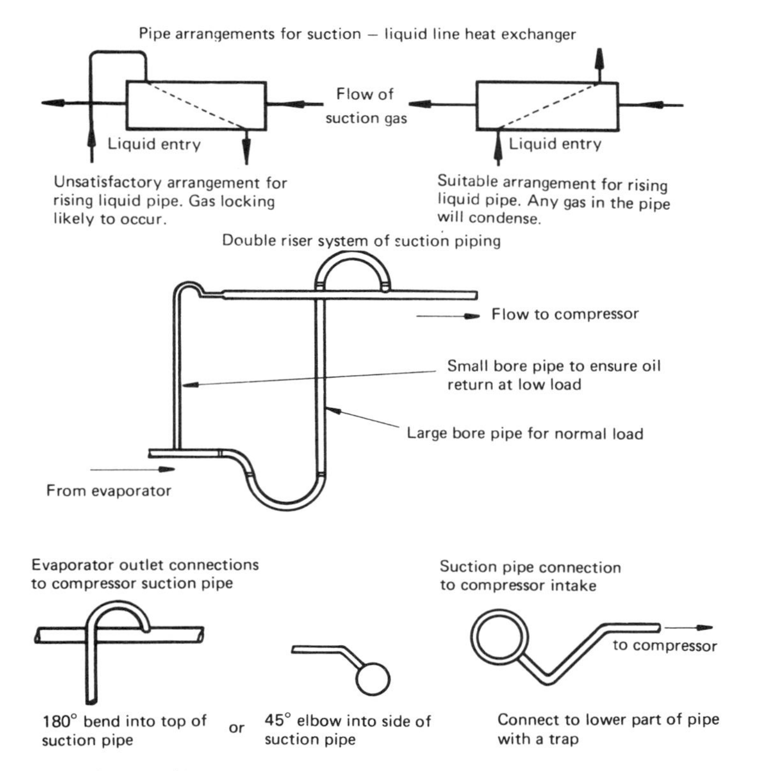

Figure 25. cont'd

6 Pipe work that has been installed should not be left open to the atmosphere longer than necessary. Plug the ends and take all possible precautions to avoid the entry of dirt or moisture.
7 Arrange the compressor delivery pipe to a remote condenser in such a way that refrigerant vapour will not condense; and return as a liquid, during the compressor off-cycle.
8 Ensure that the liquid supply pipe to the expansion valve is of adequate size. If the length is excessive or there is a 'vertical lift', use a larger than normal size. Always fit a filter-drier and sight glass just before the expansion valve and try to keep this pipe cool.
9 If it is intended that circulating oil should return to the compressor through the suction pipe, it MUST be correctly sized or seizure of the compressor may occur.

Useful hints on positioning and connecting refrigerant pipes are shown in Fig. 25.

8 PUMPING DOWN THE SYSTEM

During both installation and servicing work, it is frequently necessary to transfer all of the refrigerant charge into the high pressure side of the system. This procedure is known as 'pumping down' and is used when it is necessary to replace or repair a component fitted to the low pressure side.

1 Close the liquid receiver outlet valve. If a valve is not fitted to the system, discharge the refrigerant to the outside air and install one.
2 With all other valves in the normal running position, start the compressor and lower the pressure to slightly above atmospheric pressure, as shown on the suction gauge. DO NOT allow a vacuum to be produced as this may lead to air and other contaminants being drawn into the system.
3 Switch off the compressor and watch the suction gauge. It will usually rise as dissolved refrigerant boils out of the oil present.

When the suction pressure remains at, or slightly above, atmospheric all refrigerant has been removed from the low pressure side and repair work can begin.

Note. It may be necessary to adjust the low pressure cut out to a lower setting in order to reduce the pressure in the evaporator to 0 lb/in^2 g.

9 PURGING THE SYSTEM

This is a procedure that is carried out to remove contaminants, usually air, from the system when the time required and degree of contamination does not warrant the use of evacuation techniques.

The procedure consists of passing refrigerant vapour through the contaminated section which carries before it air, dirt, moisture etc. An outlet point is then provided through which the contaminants can be discharged.

(a) Purging using the system charge

This method is used in industrial work or when the system has valve(s) suitable for controlling the refrigerant flow. The following procedure would be adopted for purging a liquid pipe (following the replacement of a faulty component such as a filter, drier or solenoid valve) where the liquid receiver outlet was fitted with an isolating valve.

1 Remove the connection from the liquid pipe to the expansion valve inlet.
2 Open slightly the liquid receiver outlet valve. It will have been fully closed to prevent loss of refrigerant during the replacement procedure.
3 Allow the escaping vapour to discharge all contaminants from the liquid pipe to the outside air. If dirt particles are suspected ensure that the escaping vapour does not force them into the expansion valve inlet.

4 When it is considered that all contamination has been removed, remake and tighten the connection to the expansion valve. Fully open the liquid receiver outlet valve. Test the entire liquid pipe for leaks.

(b) Purging using refrigerant from a cylinder

This method is used when isolating valves are not fitted or when the quantity of refrigerant required for purging is not available in the system.

To fit a new evaporator, or repair an existing one, to a plant NOT fitted with a liquid receiver outlet valve:

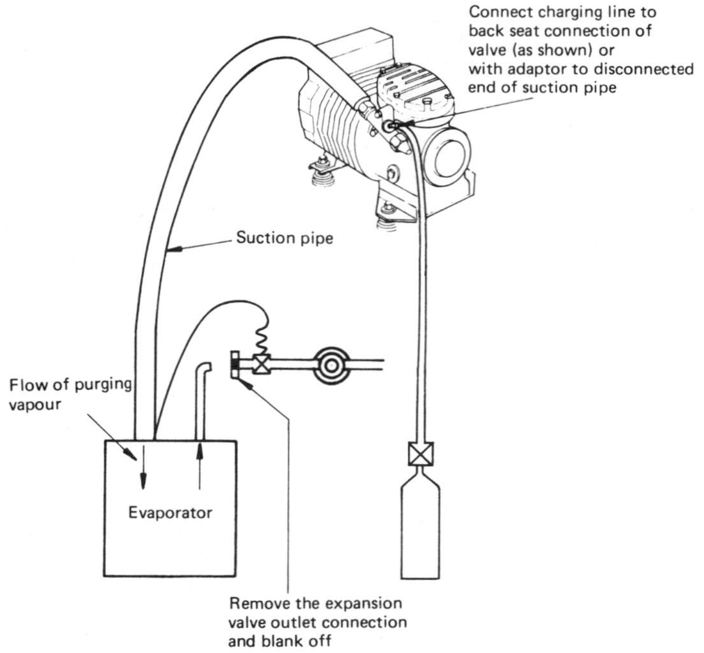

Figure 26. Purging an evaporator after repair

1 Release refrigerant from the most convenient point in the system. While the pressure is still slightly above atmospheric, isolate the high pressure side of the system as follows.
2 Remove the expansion valve outlet connection and blank off the valve with a flare nut and bonnet. Front seat the compressor suction service valve, if fitted. Otherwise, remove the suction pipe from the compressor intake and blank off with a flare nut and bonnet.
3 Upon completion of the repair, attach a charging line to the VAPOUR connection of a refrigerant cylinder. Purge this line and then connect the other end to a back seat connection of the suction service valve or, with an adaptor, to the open end of the compressor suction pipe. See Fig. 26.
4 Open the cylinder control valve sufficiently to obtain a steady flow of refrigerant through the evaporator. When it is considered that all contamination has been removed, re-connect the evaporator to the expansion valve.
5 Remove the refrigerant cylinder and charging line and re-connect the evaporator to the compressor suction. There will be sufficient refrigerant vapour present to prevent air entering while these joints are made.
6 On completion tighten all joints that have been removed or loosened, test for leaks and then re-charge the system.

10 STARTING THE PLANT

The following procedure should be followed when starting a newly installed plant for the first time, or, where applicable, starting a plant that has been shut down for a long time.

1 Ensure that the electrical phase and voltage are correct.
2 Check that the power supply cables are of sufficient capacity to handle the starting and running loads.
3 Connect the following into the compressor electrical circuit: AMMETER, VOLTMETER.
4 Unless high and low pressure gauges are fitted to the plant, install a system analyser.

5 Start the condenser fan or water circulating pump. If these are interlocked with the compressor check that they start as soon as the compressor is switched on.

6 Start the compressor with the suction pressure just above cut-in (See paragraph 8). A high suction pressure will overload the compressor motor on start up. A low suction pressure will result in oil being drawn up from the crankcase.

7 As soon as the compressor is up to normal running speed, check:

(a) The motor(s) current and voltage. The current drawn by the motor will initially be low and increase as load is applied.
(b) That the delivery head pressure is not abnormally high.
(c) That the suction back pressure is not abnormally low.
(d) That the oil pressure gauge, where fitted, indicates a satisfactory pressure.
(e) That there is an adequate flow of either air or water through the condenser.
(f) That the evaporator fan(s) or liquid chilling circulating pump(s) are running normally.

If any of the above are unsatisfactory, any component is excessively noisy or vibrates unduly, SHUT DOWN immediately. Investigate and rectify as necessary before re-starting.

8 Increase the load on the compressor slowly. With multi-evaporator circuits, admit liquid to them in turn until all are operational. One single evaporators, restrict the flow by partly closing down the liquid receiver outlet valve or the compressor suction service valve. Where these are not fitted, adjust the expansion valve to reduce the liquid flow.

The weight of vapour being drawn into the compressor MUST be low on first start and then gradually increased.

9 During the increase in load, check the items lised in paragraph 7.

Expect both the suction and discharge pressures to rise and the current being drawn by the compressor motor to increase.

Be prepared to shut down the plant if any of the unsatisfactory systems listed in paragraph 7 occur.

10 When the plant is running on full load, re-check carefully the items listed in paragraph 7 and ensure that readings taken conform to the data provided by the supplier.

11 Check the following;
   (a) That the plant is cooling properly.
   (b) That the expansion valve is correctly set.
   (c) That the controls are correctly set and that the unit cycles within the limits determined by them.
   (d) That the plant is fully charged and that there are no leaks.

12 Before handing the plant over to the customer, prepare a record (log sheet) showing operating temperatures, pressures, power consumption and cooling load.

A record of the plant's performance when new is a useful indication as to the cause of faults which subsequently occur.

11 USING A SYSTEM ANALYSER

Correct use of this component in commercial work enables the serviceman to carry out the following, during installation or servicing work, without continually making connections to the system, which takes time and increases the possibility of error and contamination.

1 Measuring high and low side pressures.
2 Charging the system with refrigerant (and discharging).
3 Adding oil to the system.
4 Evacuating the system.
5 Purging and pumping down.
6 By-passing the compressor.

The component is shown diagramatically in Fig. 27, together with instructions for carrying out the techniques 1 to 6.

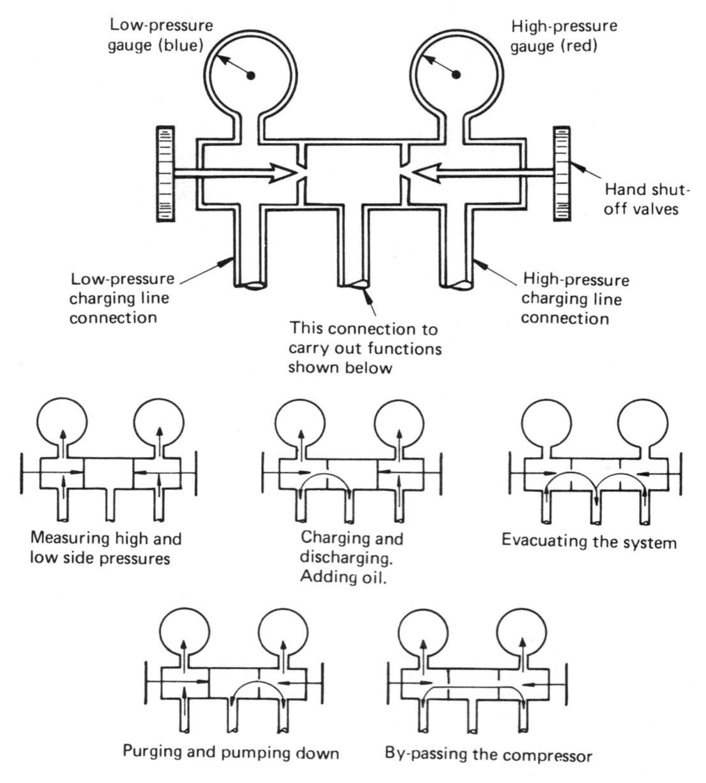

Figure 27. Use of the system analyser

12 TRANSFERRING AND HANDLING LIQUID REFRIGERANT

(a) Safety precautions

1 Handle all liquid refrigerants CAREFULLY, even the safe fluorocarbons.
2 Wear goggles and protective gloves. Do not allow liquid refrigerant to come into contact with the skin, especially the eyes.
3 Do not completely fill a cylinder. Allow room for expansion as the temperature increases. Never allow a cylinder to become more than three-quarters full.
4 Do not completely empty a cylinder, otherwise air and similar contaminants may enter.
5 Keep cylinders cool, handle them carefully, and make sure they do not get dropped or knocked.
6 Do not work in high concentrations of refrigerant, particularly AMMONIA, METHYL CHLORIDE and SULPHUR DIOXIDE.
7 When handling refrigerants ensure that escaping vapour does not come into contact with high temperatures such as electric elements and naked flames.

(b) Transferring refrigerant to a service cylinder

To transfer liquid refrigerant from a supply cylinder into a small service cylinder.

1 Remove all refrigerant vapour from the service cylinder as liquid will not flow into a cylinder containing vapour. Connect a vacuum pump to the cylinder inlet connection and reduce the pressure to between 1 and 2 Torr.
2 Weigh the cylinder carefully.
3 Reduce the temperature of the cylinder by the best means available. Spray cold water onto the outside or, preferably,

place the cylinder in a cold cabinet or deep freezer overnight.

4 Arrange a method of determining the weight of liquid admitted to the cylinder. Use a weighing machine or support the cylinder by a method that incorporates a spring balance.

5 Connect a charging line from the LIQUID outlet connection of the supply cylinder to the service cylinder inlet. Leave this connection finger tight. See Fig. 28 for the general arrangement.

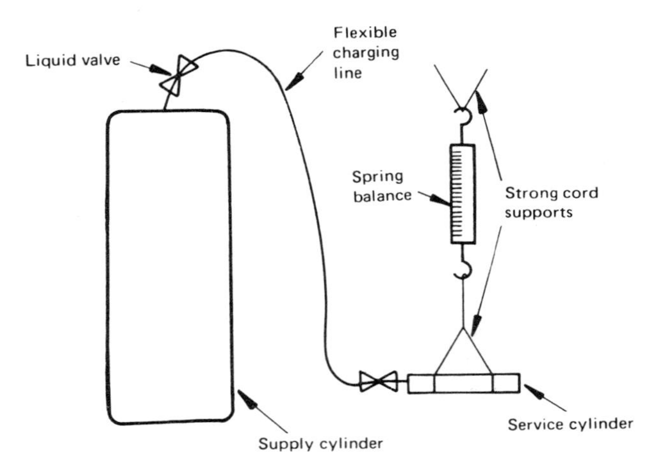

Figure 28. Transferring refrigerant into a service cylinder

6 Remove all vapour from the charging line by SLOWLY opening the supply cylinder valve and purging it out of the loose connection at paragraph 5.

7 When liquid escapes freely from this connection, tighten it. Fully open both cylinder valves and liquid will flow into the service cylinder.

8 Watch carefully the increase in weight (see paragraph 4) and close the supply cylinder valve when the service cylinder contains ¾ of its total (completely full) capacity.

9 Close the service cylinder valve and remove the charging line and weighing arrangements. Fit flare nuts and bonnets to both cylinders to prevent leakage.

10 Weigh the service cylinder, subtract the empty weight (paragraph 2) and ensure that the cylinder does not hold more liquid than recommended by the supplier, or if this is unknown, ensure that the cylinder is not more than three-quarters full.

(c) Transferring refrigerant to a graduated charging cylinder

The procedure is similar to that described in (b) above except that it is not necessary to evacuate or cool the cylinder. Vapour is removed from a purge valve fitted to the top of the cylinder.

1 Connect a charging line from the supply cylinder LIQUID outlet connection to the bottom connection of the charging cylinder. See Fig. 29. Purge this line.

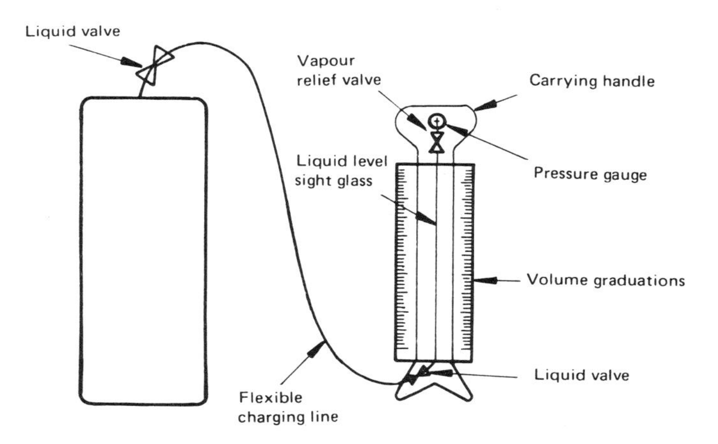

Figure 29. Transferring refrigerant into a graduated charging cylinder

2 Fully open the supply cylinder and charging cylinder valves. A small quantity of refrigerant will transfer into the charging cylinder and the pressure will be shown on the gauge mounted on the top of the charging cylinder.
3 Rotate the scale on the outside of the cylinder until the graduations for the refrigerant and its pressure are in line with the cylinder sight glass.
4 Open the purge valve, on the top of the charging cylinder, sufficiently for vapour to escape and liquid to enter. DO NOT open the purge valve more than necessary or surging will occur and, it will not be possible to read the liquid level sight glass.
5 When sufficient refrigerant has been admitted to the charging cylinder (do not charge beyond the maximum capacity graduation), close in turn the supply cylinder valve, the charging cylinder inlet valve and the purge valve. Remove the charging line and blank off both cylinders as in (b) above, to prevent leakage.

SECTION J

ELECTRICAL FAULT FINDING

1 COMPRESSOR MOTOR FAILS TO START

Listen for hum, indicating that motor is trying to start. If motor hums, proceed to 2. If there is no humming noise, check in the following sequence:

(a) Incoming supply voltage. Make sure that the supply is suitable for the motor.
(b) Supply at terminal block (distribution point). Check fuse and connections between (a) and (b) for continuity.
(c) Thermostat (stuck in open position). Place jumper lead across terminals. If motor starts, re-set or replace thermostat.
(d) High pressure cut out. Short out with jumper leads as in (c). If motor starts, check condensing pressure (See SECTION K). If pressure is normal, re-set or replace cut out.
(e) Low pressure cut out. Short out, as in (d). If motor starts, check evaporating pressure (See SECTION K). If pressure is normal, re-set or replace cut out.
(f) Motor overload protector. Short out, as above. If motor starts, check for cause of overload (See SECTION K). If no apparent cause, replace protector.
(g) Motor relay. Check electrical connections. If motor fails to start, remove relay assembly. Check solenoid continuity, operation of contacts. Observe operation of relay and replace if not functioning.
(h) Motor windings. These may be open circuit or connection detached. Test for continuity.

2 COMPRESSOR MOTOR TRIES TO START BUT DOES NOT RUN

In this condition the motor hums, when first switched on, but stops humming when the overload protector operates. Check in sequence:

(a) Low voltage. Check supply voltage with rating on motor. It must not be reduced by more than 15%. Always test voltage under load conditions i.e. at motor terminal block when the compressor is attempting to start.
(b) Motor overload protector. Connect an ammeter or watt meter to the motor circuit. Attempt to start by shorting out the overload. If the motor starts and the running current/power is normal (compared to manufacturer's data) the overload protector is faulty and needs replacing. If the motor fails to start, or the current/power is above normal, remove the jumper lead after ten seconds to avoid damage to the motor. The fault is either that the compressor motor has a locked rotor or the compressor load is too high (See SECTION K).
(c) Motor relay. Test as in 1 (g) above.
(d) Motor windings. Test as in 1 (h) above.
(e) Start capacitor. Difficult to test under field conditions. Replace.

3 COMPRESSOR MOTOR STARTS BUT DOES NOT REACH NORMAL RUNNING SPEED

In this condition, the pumping capacity of the compressor is reduced so that it runs continuously until the motor overload protector overheats and disconnects the supply. Check in sequence:

(a) Low voltage at compressor motor terminals. Test as in 2 (a).
(b) Motor relay. Test as in 2 (c).
(c) Compressor unloader faulty. Test electrical connections. For mechanical defects, see SECTION K.

(d) Excessive condensing pressure. See SECTION K.
(e) Motor windings. Test as in 2 (d).
(f) Start capacitor. Replace as in 2 (e).

4 THERMOSTAT FAILURE

May occur in either of the following conditions.

(a) STUCK OPEN. Compressor will be open-circuited and shut down. If the cold room temperature is above normal, test with jumper lead across terminals. Compressor should immediately re-start.
(b) STUCK CLOSED. Compressor may be running continuously or cycling on the overload protector. If the cold room temperature is below normal, test by turning control knob to a higher temperature. Compressor should immediately stop. Alternatively, a sharp tap on the outside will often free a stuck thermostat. Service faults (causing either (a) or (b) above).

(i) Contact points corroded or pitted. The poor electrical circuit caused by this is easily detected with a test lamp.
(ii) Connections to terminals dirty or loose. Test as in (i) above.
(iii) Loss of charge from power element. Finger pressure will be sufficient to compress bellows.
(iv) Poor thermal contact (surface control bulbs). Remove bulb from evaporator, clean, re-fit and tighten securely.

5 PRESSURE CUT-OUT FAILURE

See SECTION F CONTROLS. Service faults and diagnosis similar to those described in 4 above.

6 WIRING AND CONNECTION FAULTS

See SECTION E ELECTRICAL INSTALLATION WORK.

SECTION K

MECHANICAL FAULT FINDING

1 FAULT ANALYSIS BY TEMPERATURE AND PRESSURE

Most system faults, in both commercial and industrial plants, can be determined from the operating temperatures and pressures. A plant should normally operate at as high an evaporating pressure as possible (provided the cooling effect is being obtained) and at as low a condensing pressure as possible (provided that condensation to liquid is achieved). The operating pressures can be read from the plant gauges, or where not fitted, by use of the system analyser (see SECTION H).

Most refrigeration plants, under normal conditions, will operate at pressures corresponding to the following temperatures.

AIR COOLING EVAPORATORS	10 to 15 °F (5 to 8 °C) below cold room temperature.
LIQUID CHILLING EVAPORATORS	5 to 10 °F (3 to 5 °C) below average liquid temperature, $\frac{T_{in} + T_{out}}{2}$
AIR COOLED CONDENSERS	25 to 35 °F (15 to 20 °C) above ambient air temperature.
WATER COOLED CONDENSERS	10 to 15 °F (5 to 8 °C) above average water temperature.

Suction and delivery superheat are a further guide to determining the cause of a mechanical fault. The temperature of the vapour in the pipe above that indicated on the relevant gauge (the superheat) can be measured by attaching the probes of an electrical temperature tester, or the stems of refrigerant thermometers, (see SECTION B), to the outside of the pipe and then insulating the measurement point with expanded rubber.

Average superheat values for plants are

SUCTION PIPE	10 to 15 °F (5 to 8 °C)
DELIVERY PIPE	30 to 60 °F (17 to 33 °C) for the FLUOROCARBONS 100 to 150 °F (56 to 83 °C) for AMMONIA

Abnormal operating temperatures and pressures may be affected by the type of expansion valve fitted. In the following fault finding table, these abbreviations have been used:

EVAPORATING	EVAP
CONDENSING	COND
SUCTION	SUCT
DELIVERY	DEL
HIGH PRESSURE	HP

The causes of abnormal operating pressure and superheat have been listed in the order in which they most frequently occur. They should be examined in this order and rectified as necessary (see SECTIONS F, G and H).

For small domestic cabinets and similar, using capillary tube restrictors, this table is NOT applicable. Fault finding on these equipment is covered below under DOMESTIC SYSTEMS.

Pressure	*Superheat*	*Valve*	*Cause*
Evap. Low	Suct. High	Any type	Plant undercharged Liquid pipe obstruction Drier dirty or saturated Exp. valve faulty (closed) Liquid solenoid valve faulty (closed) Back pressure valve faulty (closed)
Evap. Low	Suct. Normal	Any type	Low evaporator cooling load Compressor not unloading Compressor suction strainer blocked
Evap. High	Suct. Low	HP float	Plant overcharged
		HP float	Valve stuck open
Evap. High	Suct. Normal	Any type	High evaporator cooling load Compressor valves leaking Back pressure valve faulty (open)
Cond. High	Del. Normal	Any type	High air/water temp. (to condenser) Condenser dirty or scaled Insufficient condenser air/water flow High evaporator cooling load Plant overcharged (except with HP float valve)
Cond. High	Del. High	Any type	Non condensable gas in system
Cond. Low	Del. High	Any type	Plant undercharged Compressor valves leaking Compressor suction strainer blocked Back pressure valve faulty (closed)
Cond. Low	Del. Normal	Any type	Low evaporator cooling load

2 METHODS OF CONFIRMING THE FAULT

It is sometimes useful to confirm the diagnosis made from the above table by making the following additional measurements.

Motor Power

Causes of high power consumption

Plant over-charged
Non-condensable gas in system
High cooling load on the evaporator
Insufficient flow of condenser cooling air/water
Condenser dirty or scaled
High temperature of air/water entering the condenser
Mechanical defects in the compressor (see ABNORMAL NOISE below)

Causes of low power consumption

Plant under-charged
Low cooling load on the evaporator
Obstruction in the liquid pipe
Drier dirty or saturated
Expansion valve faulty (closed)
Back pressure valve faulty (closed)
Compressor suction strainer blocked
Leaking suction valves

Subcooling of liquid at the condenser outlet

Average values are approximately:
for air-cooled condensers, 3 °F (1.7 °C)
for water-cooled condensers, 5 °F (2.8 °)
(if sub-cooling tubes are fitted, up to 8 °F, 4.4 °C)

Causes of excessive subcooling

Plant over-charged (except with HP float valve)

Non-condensible gas (except with HP float valve)
Obstruction in liquid pipe
Drier dirty or saturated
Expansion valve faulty (closed)
Liquid solenoid valve faulty (closed)

3 FINDING THE FAULT WHEN THE COMPRESSOR IS NOT RUNNING

Determine which control has shut down the compressor by working in the following sequence:

High pressure cut out

Short out this item (as described in SECTION F). If the compressor re-starts look for those causes of high condensing pressure listed in the table at 1 above.

Low pressure cut out

Proceed as for HIGH PRESSURE but look for causes of low evaporating pressure.

Motor overload

Test for power at the motor (as described in SECTION J). Causes of motor overload are: High Condensing Pressure (See Table). Partial seizure of compressor (See ABNORMAL NOISE below).

Differential oil pressure switch

Short out this item for a BRIEF period only. Do NOT run the compressor, simply determine that it will run. If the compressor motor attempts to start, check in the following order.

That there is sufficient oil in the compressor crankcase.
That the oil is not excessively diluted with liquid refrigerant.
That the oil filter is not blocked.
That the pressure switch is not faulty.

4 ABNORMAL NOISE PROBLEMS

Look for the cause as soon as possible. A small repair carried out early may avoid a costly replacement later.

	Probable cause
COMPRESSOR NOISY	Holding-down bolts require tightening
	Oil level too high or diluted with liquid refrigerant (oil pumping is occurring)
	Condensing pressure excessive (see paragraph 1)
	Bearings worn or valves broken
	Shipping bolts have not been removed
	Seal noisy (open compressor) low oil level or worn crankshaft seal
	Drive belts noisy. Slack and require tightening or out of alignment.
MOTOR NOISY	Holding down bolts require tightening.
	Bearings worn or require lubricating.
	Brushes worn, commutator dirty or badly worn.

MISCELLANEOUS	Noisy condensing unit. Not floating freely on mounting springs. Out of level. Base not adequately supported. Pipe work noise. Pipes require supporting to reduce vibration. Delivery pipe vibrating, fit eliminator (on large compressors fit vibration eliminators to suction and delivery connections). Pipes touching other parts of system or building.

5 DOMESTIC SYSTEM FAULTS

	Probable cause
(a) Compressor fails to start	(i) No power at motor terminals. Check incoming supply, Fuse Connections, Thermostat, Relay, Overload protector (ii) Power at Motor Terminals. Check Windings for continuity.
(b) Compressor hums but does not run	(i) Low supply voltage. Measure as compressor attempts to start. (ii) Faulty relay. Start motor with relay by-passed. (iii) Overload protector or motor windings faulty. Check as in (a). (iv) Plant is overcharged.

Fault	Cause and remedy
	Temperature test (See f below) for liquid in condenser tubes.
(c) Compressor Short cycles	(i) Electrical. Fault in wiring system, low supply voltage, relay or overload defective. (ii) Plant overcharged. During running cycle excessive frosting of suction pipe. Test as in b (iv). (iii) Condensing pressure too high. See Table 1. Confirm by measuring running current. (iv) Partial blockage of capillary tube. Check for ice (by warming) and smooth out kinks before using cleaning tool (SECTION G).
(d) Poor Cooling (long running, evaporator only partially frosted)	(i) Partial blockage of capillary tube. (See c (iv) above). (ii) Plant undercharged. Test for leaks.
(e) Poor Cooling (Evaporator de-frosts frequently)	(i) Differential setting of thermostat too great. Re-adjust or replace. (ii) Compressor cycling on motor overload protector. Low voltage or high condensing pressure. (iii) Moisture forming ice at capillary connection to evaporator which melts during off-cycle but re-forms on start up. Replace drier and re-charge unit.

Fault	Cause and remedy
(f) Poor Cooling	(i) Loss of Refrigerant charge. Warm bottom condenser tube with match or lighter. If it gets hot no liquid is present. Find leak and re-charge. (ii) Capillary tube blocked. Check pressures with system analyser. During off-cycle they should equalise in 3 to 5 mins (Domestic refrigerators) approx. 8 mins (Deep freezers).
(g) Cabinet too cold	(i) Thermostat badly positioned. Move end of capillary nearer to evaporator. (ii) Thermostat faulty. Replace. (iii) Loss of refrigerant charge leaving thermostat capillary end at higher temperature so that thermostat does not cut out. Continuous running gives low cabinet temperature. (See f (i)). (iv) Fault in 'fast freeze' system, where fitted, check switch, warning light and wiring.

SECTION L

GLOSSARY

Air conditioning	The control of temperature, humidity, purity and movement of air in an enclosed space.
Ambient temperature	The temperature of the air which surrounds the object considered.
Ampere	The unit used to measure the flow of current in an electrical circuit.
Automatic expansion valve	A pressure controlled valve designed to maintain a constant evaporating pressure.
Azeotrope	A mixture of two or more refrigerants which both evaporate and condense in a refrigerating system in the same way as a single refrigerant acting alone.
Back pressure	The evaporating pressure of a refrigerating system. The term is normally used in industrial work; alternatives are suction or low-side pressure.
Back seated	A compressor service valve turned fully in an anticlockwise direction, thus closing off the flow of vapour to the gauge connections.
Bar	A unit of pressure approximately equal to atmospheric pressure (14.7 lb/in^2).
Baudelot cooler	An evaporator used for cooling liquids, usually milk, in large industrial installations.

Bellows — A corrugated cylindrical container that is used for either providing a seal, against leakage of refrigerant, or which moves as a result of pressure change.

Bleeding — Slowly reducing the pressure of a fluid by slight opening of a suitable valve.

Brazing — A method of joining two metals with a non-ferrous filler at a temperature of approximately 800 °F (425 °C).

Calibrate — The positioning of indicators to determine accurate measurements.

Capacitor — A type of electrical storage device used in starting and/or running circuits on some types of electrical motor fitted to commercial plant.

Carrene — A name frequently given to refrigerant 11.

Cascade system — A method of producing low temperatures without the disadvantages associated with low evaporating pressure. (The evaporator of one plant is used to cool the condenser of another).

Charge — The quantity of refrigerant present in a refrigerating system.

Check valve — A valve which allows the flow of fluid in one direction only.

Circuit breaker — A safety device which automatically opens an electrical circuit that is overloaded.

Coefficient of performance — The ratio of cooling effect produced to compressor power input.

Commercial systems — Plant used for applications such as food display units, large 'walk-in' refrigerated cabinets, free-standing air cooling units and pre-fabricated cold rooms. See also 'Industrial Systems'.

Compound gauge	An instrument fitted to the evaporating side of a system used to measure pressure both above and below atmospheric.
Compound systems	Refrigerating systems that have two more compressors operating in series. Used to overcome the problems associated with high compression ratios.
Compression ratio	The ratio of the evaporating to the condensing pressure (both in units of absolute pressure).
Compressor, hermetic	A type of compressor in which the electric driving motor is sealed in the same casing as the compressor. (Used in commercial work).
Compressor, open type	A compressor in which the crankshaft extends through the crankcase. Also referred to as an external drive compressor. (Used in industrial work).
Compressor, semi-hermetic	A type of compressor in which the motor and compressor are directly coupled within the same casing but the components of which are accessible for maintenance, unlike the fully hermetic type which is completely sealed. (Used in commercial work).
Compressor seal	A seal used to prevent leakage of refrigerant where the crankshaft extends out of the crankcase in open type compressors.
Condensing unit	An assembly consisting of compressor, condenser and sometimes a liquid storage receiver mounted on a common bedplate. (Used in commercial work).
Copper plating	An electrolytic action in which copper

	is deposited on compressor surfaces.
Cracking	Opening a valve by a very small amount.
Current relay	A device used to open or close an electrical circuit. It operates by the change of current flowing in the circuit.
Cut-in	The temperature, or pressure, at which a control circuit closes.
Cut-out	The temperature, or pressure, at which a control circuit opens.
Differential	The temperature or pressure difference between cut-in and cut-out of a control.
Drier	A substance used to collect and hold moisture in a refrigerating system. Activated alumina, silica gel and molecular sieve materials are those in common use.
Dry system (direct expansion)	An evaporator from which the refrigerant is extracted as a dry vapour with no accumulation of liquid (as opposed to a flooded system).
Electrostatic filter	A type of air filter, used in air conditioning, which gives the dust particles an electric charge. These particles can then be attracted to a plate, of opposite polarity, in order that they can be removed from the air.
Epoxy resin	A synthetic plastic adhesive, used widely in evaporator repair work on commercial plant.
Evacuation	The removal of contaminating air and moisture from a refrigerating system.
External equalizer	A pipe used to connect the evaporator outlet to the underside of the bellows unit of a thermostatic expansion valve. (Used with evaporators of considerable coil length to overcome problems of friction loss).

Fail-safe control	A device in which the electrical circuit is automatically opened when the sensing element loses its pressure.
Flash gas	The instantaneous evaporation of some liquid refrigerant, which cools the remainder to the lower temperature, in an evaporator.
Flooded system	A type of refrigerating system in which liquid refrigerant fills most of the evaporator.
Flush	An operation carried out to remove contaminants from a system by purging them to the atmosphere using a refrigerant or similar fluid.
Flux	A substance applied to surfaces that are being joined by brazing or soldering to prevent oxidization.
Foaming	The formation of a foam in a refrigerant-oil mixture due to the rapid evaporation of refrigerant dissolved in the oil. This occurs when the compressor is first started and the crankcase pressure is rapidly reduced.
Freon	The trade name for a family of synthetic refrigerants used in most modern refrigerating systems.
Front seated	A compressor service valve turned fully in a clockwise direction, thus closing off the flow of vapour into and out of the compressor.
Frost back	A condition in which liquid refrigerant flows from the evaporator into the compressor suction pipe. It is usually indicated by excessive sweating or frosting on the outside of this pipe.
Fuse	An electrical safety device consisting of a strip of fusible metal which melts

	and opens the circuit when it is overloaded.
Fusible plug	A fitting made with a metal of low melting temperature which acts as a safety device to release the pressure in a system in case of fire.
Gauge manifold	A device used in servicing work which incorporates both a compound and high pressure gauge. Valves are fitted which enables the pressure at points in the system to be measured and refrigerant added.
Gauge port	An opening or connection provided for a service man to instal a gauge on a commercial plant.
Halide torch	A type of spirit or propane, burning torch used to detect refrigerant leakage from a system.
Head pressure	The condensing pressure of a refrigerating system. (Also referred to as the delivery, discharge and high-side pressure. Head pressure is the term normally used in industrial work).
High pressure gauge	An instrumentused to measure pressures above atmospheric, normally up to 400 lb/in^2. (Fitted to the condensing side of a system.)
Hot gas defrost	A system in which hot delivery gas from the compressor discharge is transferred to the evaporator at predetermined intervals, in order to remove accumulations of frost and ice.
Humidistat	An electrical control which maintains the correct humidity in an air conditioned space by varying the quantity of moisture added to or removed from the air by the air conditioning plant.

Industrial systems	Plant used for high cooling duties such as large purpose-built cold stores and ice-making installations, and the cooling requirements of industrial processes. Industrial systems normally require an attendant, whereas commercial plant is usually fully automatic.
Inhibitor	A chemical substance, used in conjunction with brine secondary refrigerants, to reduce the effects of corrosion.
Insulator, thermal	A material which is a poor conductor of heat that is used to reduce the rate of heat flow into a cold space, in order to lower the cooling capacity required of the refrigeration plant.
Latent Heat of evaporation	The quantity of heat required to be supplied to a liquid at its boiling point to change it into a vapour, or gas, without changing its temperature.
Liquid line	The pipe which carries liquid refrigerant from the condenser (or liquid received when fitted) to the expansion valve.
Liquid receiver service valve	A two or three-way service valve located at the receiver outlet and used for isolating purposes. Sometimes referred to as the King valve.
Liquor	The chemical solution used as the working fluid in an absorbtion refrigeration system.
Megger	An instrument used to test the insulation resistance of an electrical circuit.
Modulating	A type of device or control which tends to adjust by means of small changes rather than 'full on' or 'full off' operation.
Motor burn out	A condition in which the insulation of

an electric motor has been destroyed by overheating.

Muffler	A device fitted to the discharge of a compressor to reduce the noise of the pulsating gas.
Multiple system	A refrigerating system in which several evaporators are connected to one condensing unit.
Neoprene	A synthetic rubber, used in refrigeration systems for seals and joints, which is capable of resisting the corrosive action of modern refrigerants.
Neutralizer	A chemical used to counteract the action of acids that form in refrigeration systems.
Non condensable gas	A gas which will not liquefy at the plant condensing temperature and pressure.
Off cycle	The period when the compressor is not running because no further cooling is required.
Ohm	The unit used to measure resistance in an electrical circuit. A resistance of one ohm exists when an electrical potential of one volt causes a flow of one ampere.
Oil logging	Lubricating oil discharged by the compressor remaining in the evaporator, and affecting the efficiency of the plant.
Oil separator	A device fitted to the compressor discharge to remove oil from the vapour before it enters the system.
Open circuit	An interrupted circuit which prevents the flow of electricity.
Overload protector	A device, temperature, pressure or current operated, which will stop the

operation of a unit if a dangerous condition arises.

Phial — The sensing element of a thermostatic expansion valve.

Pressure drop — The difference in pressure between two points in a refrigeration system, or the two sides of a filter in an air conditioning plant.

Process tube — A length of tube attached to an hermetic compressor in order to carry out servicing work.

Pumping down — The action of using the compressor against a closed valve on the suction side to reduce the pressure.

Purging — The action of releasing refrigerant vapour from the system in order to remove contaminants from the section through which the vapour passes and discharge them to the outside air.

Skin condenser — A condenser which uses the outer surface of the cabinet for heat removal purposes.

Slugging — A condition in which a mass of liquid refrigerant or oil enters the compressor suction. The liquid, which is not compressible, generally causes hydraulic damage to the compressor.

Split system — A refrigerating or air conditioning system in which the condensing unit is installed remote from the evaporator.

Strainer — A screen or filter used to retain solid particles while allowing the free passage of liquid and vapours. Normally fitted to industrial plant.

Subcooled — Liquid refrigerant cooled below its condensing temperature but remaining at the normal condensing pressure.

Suction pressure control valve	A control, fitted in the suction pipe, which maintains a constant pressure in the evaporator during the compressor running cycle.
Superheated	Refrigerant vapour raised to a temperature above the condensing temperature but remaining at the normal condensing pressure. (Also applicable to suction vapour).
Thermostatic expansion valve	A liquid refrigerant control valve which is operated by temperature and pressure within the evaporator. A thermal phial is attached to the evaporator outlet and maintains a condition of constant superheat at this point.
Thermostatic water valve	A type of valve used to control the flow of cooling water through the shell of a water cooled condenser in order to maintain a constant condensing pressure. It also prevents water wastage by closing completely during the off cycle.
Ton of refrigeration	A rate of heat removal equal to that absorbed by ice melting at the rate of one ton per 24 hours. The unit was originally devised to help owners of cold stores decide on the size of mechanical refrigerator required when changing over from stored-ice refrigeration. It is equivalent to 12000 Btu/h or approximately 3.5 kW.
Vacuum pump	An electric motor driven pump that can be used for drawing air out of a refrigeration system. It is capable of producing the very low vacuums that are required to ensure that all moisture is removed.

Vapour barrier	A material used to prevent atmospheric water vapour penetrating an insulating material.
Vapour lock	A condition in which refrigerant vapour becomes trapped in a pipe, due to a restriction or faulty design, and prevents the flow of liquid.
Wax	An ingredient present in lubricating oils which separates from the oil at low temperature and causes blockages in the system.
Wire drawing	The resistance to vapour flow as it passes through an orifice or across the seating of a valve. At high vapour velocity the effect of wiredrawing is to score and damage the valve seat.

SECTION M

METRIC CONVERSION TABLE

	To convert from	*To*	*Multiply by*
Pressure	atmospheres	bars	1.013
		kilograms per sq. cm	1.033
		pounds per sq. inch	14.696
	bars	atmospheres	0.987
		kilograms per sq. cm	1.020
		pounds per sq. inch	14.504
	kilograms per sq. cm	atmospheres	0.968
		bars	0.981
		pounds per sq. inch	14.223
	pounds per sq. inch	atmospheres	0.0680
		bars	0.0689
		kilograms per sq. cm	0.0703

Note: 1 bar = 100 kN/m^2 = 100 kPa

Power	tons of refrigeration	btu per min	200
		kilowatts	3.517
		kilocalories per min.	50.4
	btu per min	tons of refrigeration	0.005
		kilowatts	0.01758
		kilocalories per min.	0.2520
	kilowatts	tons of refrigeration	0.2843
		kilocalories per min.	14.330
		btu per min.	56.869
	kilocalories per min	tons of refrigeration	0.0198
		btu per min	3.968
		kilowatts	0.0698
Length	inches	millimetres (mm)	25.40
	millimetres (mm)	inches	0.03937
Temperature	°F	°C	°C = (°F − 32) $\frac{5}{9}$
	°C	°F	°F = (°C × $\frac{9}{5}$) + 32

INDEX